Akira Hasegawa

Optical Solitons in Fibers

Akira Hasegawa

Optical Solitons in Fibers

Second Enlarged Edition

With 25 Figures

Springer-Verlag Berlin Heidelberg GmbH

Professor Dr. Akira Hasegawa
AT & T Bell Laboratories, 600 Mountain Avenue
Murray Hill, NJ 07974-2070, USA

The first edition appeared in the series
Springer Tracts in Modern Physics, Vol. 116

DOI 10.1007/978-3-662-09113-5

Library of Congress Cataloging-in-Publication Data. Hasegawa, Akira, 1934 –. Optical solitons in fibers / Akira Hasegawa. – 2nd enl. ed. p. cm. Includes bibliographical references. 1. Solitons. 2. Optical fibers. I. Title. QC174.26.W28H37 1990 530.4'1–dc20 89-26313

Originally published by Springer-Verlag Berlin Heidelberg New York in 1990
MyCopy version of the original edition 1990

Cover design: W. Eisenschink, D-6805 Heddesheim

2157/3150-543210 – Printed on acid-free paper
www.springer.com/mycopy

To *Professor Tosiya Taniuti*
who introduced the author to the concept of solitons

Preface to the Second Edition

The first edition of this booklet found its readers rather quickly, making it necessary to prepare this second updated edition. In view of the fact that only a few months had elapsed, it was only desirable to make some minor corrections and to include new material on the most recent experimental results on erbium doped fibers (Sect. 6.4) and on dark solitons (Sect. 10.2). There is little to add in respect to theory and to the related development in present and future applications of new technologies in the distortionless signal transmission in ultra-high speed telecommunications.

Since the book is suitable for natural scientists in general and for graduate or last year undergraduate courses, it was decided to produce this second edition as a softcover version with a lower price, thus making it easier for students to acquire their personal copy.

Murray Hill, November 1989 *A. Hasegawa*

Preface to the First Edition

The word *soliton* was coined by Zabusky and Kruskal in 1964 when they discussed the particle-like behavior of numerical solutions of the Korteweg deVries equation. The solitons emerge unchanged from collisions with each other and regain their asymptotic shapes, magnitudes, and speeds. In 1967 Gardener et al. discovered that the Korteweg deVries equation is analytically soluble and that its intrinsic solutions are solitons. Solitons are now fundamental objects with uses in a wide variety of applications in various branches of science.

In 1973 Hasegawa and Tappert showed theoretically that an optical pulse in a dielectric fiber forms an envelope soliton, and in 1980 Mollenauer demonstrated the effect experimentally. This discovery is significant in its application to optical communications where conventional optical pulses are distorted due to fiber dispersion.

Optical solitons are currently under intense study since they are expected to provide not only distortionless signal transmission in ultra-high speed telecommunications but also many interesting nonlinear optical applications in fibers.

This book gives a concise yet readily accessible introduction to the concept of optical solitons and to their use in signal propagation through fibers. It gives a clear presentation of both the basic theoretical connections and the experimental aspects and includes a discussion of the most recent developments.

The text will be useful not only for researchers and graduate students but also for the lecturer who wants to introduce this fascinating and novel concept to undergraduate or graduate students.

Murray Hill, May 1989 *A. Hasegawa*

Contents

1. Introduction

It is approximately a quarter of a century since the word "soliton" was first introduced. During this period, various types of solitons have been discovered in almost all areas of physics. Approximately one hundred different types of nonlinear partial differential equations have been found to have soliton, or soliton-like solutions mathematically. The definition of a soliton has also expanded from a narrowly defined one which refers to a localized solution of an exactly integrable solution of a nonlinear partial differential equation to a more general one which refers to a relatively stable nonlinear solitary wave, even if the equation is not integrable. Various definitions have been found to be convenient, depending on the physical problem to be solved.

The word "soliton" was first used in a paper by *Zabusky* and *Kruskal* [1.1] published in 1964. In this paper, the Korteweg–deVries (KdV) equation [1.2], which had been known since the 19th century, was solved numerically for a periodic boundary condition in order to model the one-dimensional nonlinear oscillation of a lattice. The authors discovered that a train of solitary waves was formed, and that these solitary waves passed through one another without deformations due to collisions. Consequently, they named these solitary waves "solitons", since they behaved like particles.

The Korteweg–deVries equation was derived in the 19th century to model a wave which propagates on the surface of water. The fact that the equation has a solitary wave solution has, in fact, been known since that time. Zabusky and Kruskal discovered numerically that these solitary waves were *naturally formed* from a periodic (sinusoidal) initial condition and that they were *stable* when they collided with one another. In this respect, the soliton is a product of the high-speed computer technology of the 20th century. Two years later, *Gardner* et al.[1.3] succeeded in giving the mathematical interpretation of these solitary wave solutions. Using the inverse-scattering method which was originally developed for wave functions in quantum–mechanics, they showed that the KdV equation can be solved exactly for a localized initial condition. From this solution, it was discovered that soliton solutions correspond to the bound state of a Schrödinger operator and that consequently, the particle picture of the solitary waves could be mathematically verified.

The optical soliton which will be treated in this book differs somewhat from KdV solitons. In contrast to the KdV soliton which describes the solitary wave of a wave, the optical soliton in fibers is the solitary wave of an envelope of a light wave. In this respect, the optical soliton in a fiber belongs to the category which is generally referred to as an envelope soliton. The optical pulses which are used in communication are ordinarily created by the pulse modulation of a light wave. Here also, the pulse shape represents an envelope of a light wave.

The fact that an envelope of a wave which propagates in a strongly dispersive nonlinear medium, that is to say a medium in which the wave property depends on the amplitude as well as on the wavelength of the wave, has a solitary wave solution had been known for waves in plasmas [1.4,5] since around the same time as the word "soliton" was introduced. It was also known that the localized envelope wave has a relation akin to the phenomena of the self-focusing of light [1.6], which were studied experimentally during this time.

The model equation which describes the envelope soliton propagation is known as the nonlinear, or cubic Schrödinger equation and, using the complex amplitude $q(x,t)$, is expressed by

$$\mathrm{i}\,\frac{\partial q}{\partial x}+\frac{1}{2}\,\frac{\partial^2 q}{\partial t^2}+|q|^2 q=0 \quad .$$

In this equation, x represents the distance along the direction of propagation, and t represents the time (in the group velocity frame). The second term originates from the dispersion of the group velocity, i.e., the fact that the group velocity is dependent on the wavelength, and the third term originates from the nonlinear effect, i.e., that fact that the wavelength depends on the intensity of the wave.

Hasegawa and *Tappert* [1.7] were the first to show theoretically that an optical pulse in a dielectric fiber forms a solitary wave due to the fact that the wave envelope satisfies the nonlinear Schrödinger equation.

However, at that time, neither a dielectric fiber which has small loss, nor a laser which emits a light wave at the appropriate wavelength ($\simeq 1.5\ \mu$m), was available. Furthermore, the dispersion property of the fiber was not known and, consequently, it was necessary to consider a case where the group dispersion $\partial^2 k/\partial\omega^2$ is positive, that is to say when the coefficient of the second term in the nonlinear Schrödinger equation is negative. Here, k and ω represent the wave number and angular frequency of the light wave, respectively. *Hasegawa* and *Tappert* [1.8] have also shown that, in the case where the coefficient of the second term is negative, the solitary wave appears as the absence of a light wave, and called this a "dark" soliton. One year prior to the publication of the paper by Hasegawa and Tappert, *Zakharov* and *Shabat* [1.9] showed that the nonlinear Schrödinger equation can be solved using the inverse scattering method, in a similar way as previously in the case of the KdV equation. According to this theory, the properties of the envelope soliton of the nonlinear Schrödinger equation can be described by the complex eigenvalues of Dirac-type equations, the potential being given by the initial envelope wave form.

The origin of the nonlinear property in an optical fiber is the Kerr effect [1.10–12] which produces a change in the refractive index of glass owing to the deformation of the electron orbits in glass molecules due to the electric field of light. The Kerr coefficient n_2, which represents a change in the refractive index from n_0 to $n_0+n_2E^2$ for the electric field E of light, has a very small value of 10^{-22} (m/V)^2. Although the electric field in a fiber has a relatively large magnitude ($\simeq 10^6$ V/m) (for an optical power of a few hundred mW in a fiber with a cross section of 100 μm^2), the total change in the refractive index is still 10^{-10}, and this seems to be negligibly small. The reason why such a small change in the refractive index becomes important is that the modulation frequency $\Delta\omega$, which is determined by the inverse of the pulse width, is much smaller than the frequency of the wave ω and secondly, the group velocity dispersion, which is produced by $\Delta\omega$, is also small. For example, the angular frequency ω for the light wave with wavelengths λ of 1.5 μm is $1.2 \times 10^{15}\,s^{-1}$. If a pulse modulation with a pulse width of 10 ps is applied to this light wave, the ratio of the modulation frequency $\Delta\omega$ to the carrier frequency ω becomes appropriately 10^{-4}. As will be shown later, the amount of wave distortion due to the group velocity dispersion is proportional to $(\Delta\omega/\omega)^2$ times the coefficient of the group dispersion k'' ($= \partial^2 k/\partial\omega^2$). Consequently, if the group dispersion coefficient is of the order of 10^{-2}, the relative change in the wave number due to the group dispersion becomes comparable to the nonlinear change.

However, in order for these effects to become significant, the wave distortion due to the fiber loss should be less than these small effects. This requires that the fiber loss rate per wavelength of light should become less that 10^{-10}. Technically, this means that the power loss rate of the fiber should be less than 1 dB per km. Because of this requirement, seven years elapsed between the publication of Hasegawa and Tappert and the first experimental demonstration of an optical soliton transmission. At that time, a fiber with sufficiently small loss and a color center laser with a tunable wavelength appeared. In 1980, *Mollenauer* et al. [1.13] propagated a 10 ps optical pulse with a wavelength of 1.5 μm and a peak power of a few W through a 700 meter low loss fiber and, by showing that the pulse width contracts with an increase in the peak power, demonstrated for the first time the successful propagation of optical solitons in a fiber.

The fact that the envelope of a light wave in a fiber can be described by the nonlinear Schrödinger equation can also be illustrated by the generation of modulational instability when a light wave with constant amplitude propagates through a fiber [1.14–16]. Modulational instability is a result of the increase in the modulation amplitude when a wave with constant amplitude propagates through a nonlinear dispersive medium with anomalous dispersion, $k'' < 0$. The origin of modulational instability can be identified by looking at the structure of the nonlinear Schrödinger equation. If we consider that the third term in the nonlinear Schrödinger equation represents the equivalent potential which traps the quasi-particle described by the Schrödinger equation, the fact that the potential is proportional to the absolute square of the wave function indicates that the potential depth becomes deeper in proportion to the density of the quasi-particle. Consequently, when the local density of the quasi-particle increases, the trapping potential increases further, thus enhancing the self-induced increase of the quasi particle density. This process leads to modulational instability. The first experimental verification of the modulational instability of the light wave in a fiber was demonstrated by *Tai* et al. [1.17] in 1986.

The excitation of dark solitons in wavelengths greater than 1.3 μm, where the group dispersion becomes positive, was also demonstrated recently by *Emplit* et al. and *Krökel* et al. [1.18]. Thus, we see that the validity of describing the envelope of a light wave in a fiber by the nonlinear Schrödinger equation has been demonstrated in many examples.

Many applications of optical solitons can be considered. One important possibility is their application in optical communications. The optical communication technology which is already in practical use utilizes an optical pulse train with a pulse width of approximately one nanosecond. When the pulse width is this large, the major distortion of the pulse originates from the fiber loss. In order to correct this distortion, a repeater is installed at every several tens of kilometers. However, when the pulse width is decreased in order to increase the transmission rate to the level of 10 picoseconds, the separation between two repeaters is decided not by the fiber loss, but by the group velocity dispersion of the pulse. In this regime, the distance between two repeaters becomes shorter in inverse proportion to the square of the pulse width. In an attempt to overcome this difficulty, a fiber which has small group dispersion is being developed.

On the other hand, an optical soliton which is produced by the balance between the nonlinear effect and the group dispersion effect produces no distortion as a consequence of the dispersion. However, when the light intensity of the soliton decreases due to the fiber loss, the pulse width of the soliton expands. Because of this, a soliton transmission system also requires that the pulse be reshaped. However, in contrast to the linear system, the reshaping can be achieved utilizing only optical amplifiers.

Fortunately, if one utilizes the Raman effect in the fiber, optical solitons can be continuously amplified by a pump light wave which is transmitted simultaneously through the fiber. When the Raman gain is adjusted so as to compensate for the fiber loss, an optical soliton can propagate without distortion over an extremely long distance [1.19]. Recently, *Mollenauer* and *Smith* [1.20] succeeded in demonstrating a distortionless transmission of 50 ps soliton pulses over a distance of 6000 kms by utilizing this method. This demonstration has confirmed the feasibility of all-optical transmission systems which do not require repeaters.

Another example of an optical soliton application is the soliton laser invented by *Mollenauer* [1.21]. By utilizing the fiber as part of the cavity of a laser, the laser action can be controlled by the soliton which propagates through the fiber. Solitons with a pulse of the order of one picosecond have been generated by this method.

One can also utilize the soliton process to compress an optical pulse. This exploits the fact that the pulse contracts itself by the Kerr effect when the intensity of the input pulse is much larger than the intensity required to form a soliton. A pulse of the order of several tens of femtosecond was successfully produced by this method [1.22]. It is worth pointing out that these applications of solitons have become available only a quarter of a century after the birth of the soliton concept.

A further interesting fact regarding the nature of an optical soliton in a fiber is the observation of the effect of higher order terms which cannot be described by the nonlinear Schrödinger equation. The small parameter used for the derivation of the nonlinear Schrödinger equation is of the order of 10^{-5}. Therefore, the description of a process which is of higher order in this small parameter would seem to require an extremely accurate experiment.

An example of this higher order effect is the Raman process which exists within the spectrum of a soliton. When the central spectrum of a soliton acts as a Raman pump, amplifying the lower sideband spectra within the soliton spectra, the frequency spectrum gradually shifts to the lower frequency side. This effect was first observed in an experiment by *Mitschke* and *Mollenauer* [1.23], and was theoretically explained by *Gordon* [1.24] in terms of the induced Raman process. *Kodama* and *Hasegawa* [1.25] have identified that Mollenauer's discovery is due to the higher order term which represents the Raman effect. It is thus shown that the Raman process in the spectrum of a soliton produces a continuous shift of the soliton frequency spectrum to the lower frequency side, without changing the pulse shape.

The self-induced Raman process can be utilized to split two or more solitons which are superimposed in a fiber. The splitting of a soliton pulse produced by the Raman process was recently demonstrated by the experiments of *Beaud* et al. [1.26] and of *Tai* et al. [1.27]. The reason why it is possible to detect a process which is dependent on such a small parameter is that the light frequency, which is of order 10^{14}, is very large and, therefore, even if the perturbation is of order 10^{-10}, the modification in the light frequency becomes 10^4Hz and is consequently readily observable.

The series of successful experimental and theoretical studies of the behaviour of optical solitons have now lead many people to seriously consider using solitons in optical communications [1.28–31]. Optical solitons in fibers are now becoming "a real thing", rather than merely interesting mathematical objects.

This book describes the behaviour of optical solitons in fibers starting from the introduction of elementary processes of nonlinear wave propagation and by deriving the nonlinear wave equation for a wave envelope and by describing the soliton concept. It also introduces various experimental results of light wave in fibers which demonstrate the soliton and soliton related properties.

2. Wave Motion

In this chapter we present mathematical descriptions of wave motions in a nonlinear dispersive medium, and in a glass fiber.

2.1 What is Wave Motion?

In order to understand solitons, we have to review what is known about wave motion. The wave which most people first encounter may be one which approaches a beach. Let us attempt to describe mathematically wave motions of this type. The wave on a beach is a phenomenon of the motion of the water surface moving up and down. Hence, in order to describe the wave quantitatively, it is convenient to use the height of the water surface. When we enter the water and stand still, we observe that the height of the wave, h, which is the position of the surface of the water, moves periodically up and down, say, from feet to breast. This means that the quantity h is a periodic function of time when it is observed from a fixed point. We describe this periodic motion in time by

$$h = h_0 \cos \omega t \quad . \tag{2.1}$$

Here, ω is the angular frequency of the periodic motion, and h_0 the amplitude.

Let us now stand on the beach once more and observe the wave. We also note that the waves approach the beach without significantly changing their shape. When our eyes follow the crest of the wave, the shape of the wave does not change in time. If we describe the speed of the wave motion by v, then the coordinate ξ, which moves with the wave, can be described by the expression

$$\xi = x - vt \quad , \tag{2.2}$$

where x is the coordinate fixed to the frame of reference, which is the beach. The fact that the shape of the wave does not vary for an observer in the ξ-system moving with the wave, indicates that the height of the wave h can be described by a function of the coordinate ξ alone, without the time t.

We note that the wave also behaves periodically in the coordinate ξ (see Fig. 2.1). If we write this periodic function as

$$h(\xi) = h_0 \cos k\xi \quad , \tag{2.3}$$

the quantity k represents the periodicity in the ξ coordinate, and is called the

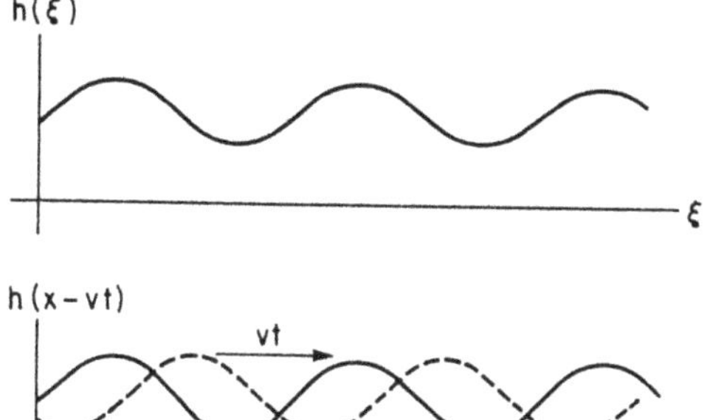

Fig. 2.1. Wave motion $h(\xi)$ and $h(x-vt)$ of water surface. ξ is the coordinate which moves with the wave phase velocity

wave number. The wavelength λ and wave number k have the relation

$$k = \frac{2\pi}{\lambda} \quad . \tag{2.4}$$

This relation can be understood as the fact that the phase of the wave changes from 0 to 2π when ξ changes from 0 to λ.

If we describe ξ in terms of the stationary coordinate x, using (2.2), (2.3) becomes

$$\begin{aligned} h(x,t) &= h_0 \cos k(x-vt) \\ &= h_0 \cos(kx-kvt) \quad . \end{aligned} \tag{2.5}$$

If we compare (2.5) with (2.1), we see that

$$kv = \omega \quad . \tag{2.6}$$

From this, the quantity $v(=\omega/k)$ is called the phase velocity of the wave. In general, the wave number k is a vector designating the direction of the wave propagation $\boldsymbol{k}$, and is related to the vector phase velocity through $\boldsymbol{v} = (\omega/k^2)\boldsymbol{k}$. The vector $\boldsymbol{k}$ is called the wave vector.

2.2 Dispersive and Nonlinear Effects of a Wave

Let us return to the beach and observe the wave more carefully. We note that, as the wave approaches the beach, the shape of the wave changes gradually from sinusoidal to triangular. This is a consequence of the nonlinear nature of the wave, where by the crest of the wave moves faster than the rest (see Fig. 2.2). As the wave approaches the beach, this nonlinear effect is enhanced and, consequently, the wave breaks up near the beach. Since the speed of wave propagation depends on the height of the wave, this phenomenon is termed a nonlinear effect. If the wave phase velocity, v, depends weakly on the height of the wave, h, (2.6) becomes

$$v = \frac{\omega}{k} = v_0 + \delta_1 h \quad , \tag{2.7}$$

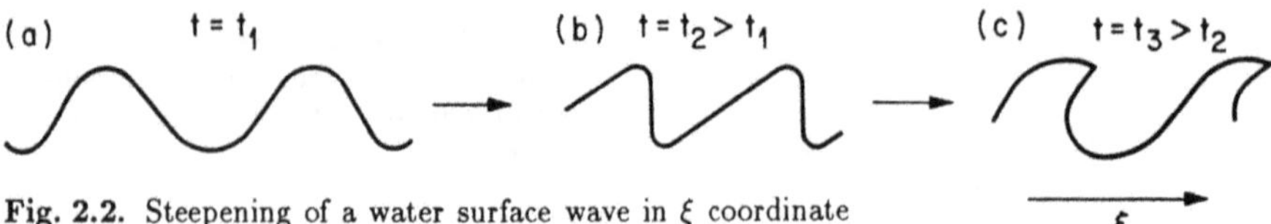

Fig. 2.2. Steepening of a water surface wave in ξ coordinate due to nonlinearity

where $\delta_1 = \partial v/\partial h|_{h=h_0}$, h_0 is the average height of the water surface, v_0 is a linear portion of the wave phase velocity (i.e., the phase velocity which does not depend on the height of the wave), and δ_1 is a coefficient representing the nonlinear effect.

In addition to the nonlinear effect, a wave also has a dispersive effect. This is an effect in which the wave phase velocity depends on the wave frequency, or wave number. When the dispersive effect is weak, it can be described as a small deviation from the dispersionless phase velocity v_0 in (2.7),

$$v = \frac{\omega}{k} = v_0 + \delta_2 k^2 \quad , \tag{2.8}$$

where $\delta_2 = \partial v/\partial k^2|_{k^2=0}$. Here, δ_2 represents the coefficient of the dispersion property of the wave. Normally, the lowest order dispersion appearing in the phase velocity is proportional to k^2. The term proportional to k represents dissipation.

2.3 Solitary Waves and the Korteweg deVries Equation

In the absence of dispersion and nonlinearity, a wave which propagates in one direction can be described by the coordinate ξ, as shown in (2.3). This indicates that, in the coordinate moving with the wave ξ, the wave can be described by

$$\frac{\partial h}{\partial \tau} = 0 \quad . \tag{2.9}$$

Here, τ represents the time in this coordinate,

$$\xi = x - vt \tag{2.2}$$

$$\tau = t \quad . \tag{2.10}$$

If we turn to the original coordinate $x = \xi + v\tau$, $t = \tau$, (2.9) becomes

$$\frac{\partial h}{\partial t} + v\frac{\partial h}{\partial x} = 0 \quad , \tag{2.11}$$

where use is made of the fact that

$$\frac{\partial}{\partial \tau} = \frac{\partial}{\partial t}\frac{\partial t}{\partial \tau} + \frac{\partial}{\partial x}\frac{\partial x}{\partial \tau} = \frac{\partial}{\partial t} + v\frac{\partial}{\partial x} \quad . \tag{2.12}$$

Let us now introduce the nonlinear effect which modifies the phase velocity, v, as a function of the wave height h. From (2.7), $v = v_0 + \delta_1 h$. Hence, the nonlinear effect introduces an additional term which is proportional to $\delta_1 h(\partial h/\partial x)$ in (2.11). This means that in ξ, τ coordinates, (2.9) becomes

$$\frac{\partial h}{\partial \tau} + \delta_1 h \frac{\partial h}{\partial \xi} = 0 \quad . \tag{2.13}$$

This expression indicates that the wave height, h, can no longer be a function of ξ alone, but also becomes a function of τ. This means that even if we move with the coordinate traveling at the linear phase velocity v_0, the quantity, h, varies both in time and space.

Let us now consider the effect of phase velocity dispersion. If the phase velocity deviates from the linear phase velocity by a quantity which is proportional to k^2, as shown in (2.8), the wave frequency $\omega (= kv)$ deviates by $\delta_2 k^3 v_0$. k^3 corresponds to the third derivative with respect to ξ in the ξ space. Consequently, if we include the effect of both linear and nonlinear velocity dispersion, the wave motion in the $\xi - \tau$ space can be described by

$$\frac{\partial h}{\partial \tau} + \delta_1 h \frac{\partial h}{\partial \xi} + \delta_2 \frac{\partial^3 h}{\partial \xi^3} = 0 \quad . \tag{2.14}$$

If we normalize the quantities ξ and h so that δ_1 and δ_2 become unity, (2.14) takes the form

$$\frac{\partial h}{\partial \tau} + h \frac{\partial h}{\partial \xi} + \frac{\partial^3 h}{\partial \xi^3} = 0 \quad . \tag{2.14$'$}$$

Equation (2.14′) is called the Korteweg deVries (KdV) equation and is, in fact, known to describe wave motion on the surface of shallow water.

The KdV equation has a structure in which the dispersive and nonlinear terms can balance to form a stationary solution. This is because, when the wave becomes steeper and triangular-like in shape as it approaches the beach, the dispersion effect (that is, the third derivative of h with respect of ξ) comes into play at the triangular corner tip of the wave and produces a smoothing effect at that corner. In fact, (2.14′) has a solitary wave solution,

$$h(\tau, \xi) = 3\eta \operatorname{sech}^2 \frac{\sqrt{\eta}}{2} (\xi - \eta\tau) \quad , \tag{2.15}$$

by the balance of the second and the third term. The solitary wave solution is stationary in the coordinate moving at a speed η in the $\xi - \tau$ space. This means that the wave moves at a speed of $\eta + v_0$ in the stationary frame. The reader may wish to confirm that (2.15) does, in fact, satisfy the KdV equation.

A unique property of this solution is that the wave height 3η, the width of the solitary wave $2/\sqrt{\eta}$, and the speed of the wave η in the $\xi - \tau$ space are related to each other through one common parameter, η. The solitary wave with a greater height, h, has a faster speed and smaller width. As will be shown in

Sect. 2.4, Gardner et al. have shown that the KdV equation is exactly soluble for a localized initial condition and that the solution can be described by a group of solitary waves, the amplitudes η described in (2.15) being the eigenvalues of a linear Schrödinger equation with the potential given by the initial condition. This fact shows that the solitary wave solution of (2.15) is not only a particular solution of the KdV equation, but that it is also a characteristic solution of the nonlinear wave. This indicates that solitons play a role similar to the Fourier modes in a linear system. This fact is also the origin of the stability of solitons when undergoing collisions.

2.4 Solution of the Korteweg deVries Equation

We introduce here a means of solving the KdV equation using the inverse scattering method [2.1]. Following *Gardner* et al. [2.1], we write the KdV equation in the form

$$\frac{\partial u}{\partial t} - 6u\,\frac{\partial u}{\partial x} + \frac{\partial^3 u}{\partial x^3} = 0 \quad . \tag{2.16}$$

Gardner et al. have considered the Schrödinger-type eigenvalue problem with the potential given by the solutions of the KdV equation (2.16)

$$\frac{\partial^2 \phi}{\partial x^2} + [\lambda - u(x,t)]\,\phi = 0 \quad . \tag{2.17}$$

It can then be shown that if $u(x,t)$ satisfies the KdV equation, the eigenvalue λ is independent of time, and the equation which describes the time evolution of $\phi(x,t)$ is given by

$$\frac{\partial \phi}{\partial t} = -4\,\frac{\partial^3 \phi}{\partial x^3} + 3\,\frac{\partial u}{\partial x}\,\phi + 6u\,\frac{\partial \phi}{\partial x} \quad . \tag{2.18}$$

Finding an eigenvalue equation which satisfies this property, i.e., the eigenvalue becomes independent of time when the potential satisfies certain nonlinear time evolution equations, has now become a common feature of methods of finding soliton solutions for the nonlinear time evolution equation. *Lax* [2.2] discovered a method of constructing the eigenvalue equation which satisfies this property. The Lax method was later used by *Zakharov* and *Shabat* [2.3] to solve the nonlinear Schrödinger equation.

Let us now return to (2.17) and consider how to solve the KdV equation. We first note that the fact that the eigenvalue λ remains invariant in time means that if λ is given for the initial value of $u(x,t=0)$, it remains the same for any time t when $u(x,0)$ evolves to $u(x,t)$. For a given potential $u(x,0)$, (2.17) can be solved for the wave function ϕ with wave number k at $|x| \to \infty$. This problem is called the scattering of a wave by the potential $u(x,0)$. By solving the scattering problem, one can obtain scattering data such as the transmission coefficient $1/a(k)$, the reflection coefficient $b(k)/a(k)$, the eigenvalues $\lambda_n(= -\kappa_n^2)$ and the normalization coefficients c_n $(n = 1, 2, ..., N)$ of the eigenfunction.

The time evolution of the scattering data can now be obtained as

$$\left.\begin{aligned} a(k,t) &= a(k,0) \\ b(k,t) &= b(k,0)\,\exp(8ik^3t) \\ c_n(t) &= c_n(0)\,\exp(4\kappa_n^3 t) \end{aligned}\right\} \quad , \tag{2.19}$$

using the fact that $u \to 0$ as $|x| \to \infty$.

The solution of the KdV equation $u(x,t)$ is then obtained using the inverse scattering method of obtaining the potential $u(x,t)$ of the Schrödinger equation (2.17) for the given time-dependent scattering data of (2.19). The inverse scattering method, which is well-established in quantum mechanics, gives

$$u(x,t) = -2\,\frac{\partial}{\partial x}\,K(x,x;t) \quad , \tag{2.20}$$

where K is given by the solution of the linear integral equation,

$$K(x,y;t) + F(x+y;t) + \int_{-\infty}^{x} K(x,z;t)\,F(z+y;t)\,dz = 0 \quad , \tag{2.21}$$

the kernel F being given by the scattering data,

$$F(x;t) = \sum_{n=1}^{\infty} c_n^2(t)\,\mathrm{e}^{-\kappa_n x} + \frac{1}{2\pi}\int_{-\infty}^{\infty} \frac{b(k,0)}{a(k,0)}\,\mathrm{e}^{ikx}\,dk \quad . \tag{2.22}$$

In order to illustrate how the soliton solution originates from the inverse scattering method, let us consider a case where the initial value $u(x,0)$ of the KdV equation corresponds to a simple potential such that it has only one eigenvalue ($N = 1$), and a zero reflection coefficient $b(k,0) = 0$. Then, from (2.19) and (2.22), F is given by

$$F(x;t) = c(0)\,\mathrm{e}^{-\kappa x + 8\kappa^3 t} \tag{2.23}$$

and the solution of the integral equation (2.21) is obtained using the Fourier transformation,

$$K(x,y;t) = \frac{c^2(0)\,\mathrm{e}^{-\kappa(x+y)+8\kappa^3 t}}{1 + c^2(0)\,\mathrm{e}^{-\kappa x + 8\kappa^3 t}/(2\kappa)} \tag{2.24}$$

where $u(x,t)$ is given by (2.20),

$$u(x,t) = -2\kappa^2\,\mathrm{sech}^2[\kappa(x - 4\kappa^2 t) - \delta] \tag{2.25}$$

with

$$\delta = \frac{1}{2}\,\ln\left[\frac{c^2(0)}{2\kappa}\right] \quad , \tag{2.26}$$

which is the soliton solution. The above derivation shows that the soliton is the unique solution of the localized initial value of $u(x, t = 0)$, which satisfies the particular property of the potential prescribed in this example. The single parameter κ, which describes the soliton solution, is given by the time-invariant eigenvalue of (2.17).

The above example can be extended to the more general case of a localized initial condition $u(x, 0)$ which gives a potential of N-bound states with no reflection of the wave function ϕ of (2.17). The N-soliton solution then emerges as the solution of (2.20) and (2.22), their parameters κ_n being given by the invariant eigenvalues. Therefore, soliton solutions are unique solutions of the KdV equation for such a localized initial condition. Furthermore, the stationary nature of the eigenvalues provides the important property of the stability of the soliton when undergoing collisions. Hence, solitons are important not only as a particular solution of the KdV equation, but as a unique solution whose stability is guaranteed by the time-invariant property of the eigenvalue of (2.17). The fact that the solution of the KdV equation for a localized initial condition can be expressed in terms of N-sets of solitons indicates that solitons can be identified as fundamental to the nonlinear equation, playing the role of the Fourier modes in a linear system.

3. Envelope of a Light Wave

In this chapter we describe a mathematical formulation of a wave envelope in a nonlinear dispersive medium, and in a glass fiber.

3.1 Envelope of a Wave

Let us return to the beach again and watch the wave approaching the shore. We recognize that waves do not necessarily have the same amplitude. If we observe them carefully, we note that several waves with relatively low amplitude are followed by one or two waves with large amplitude. Hence, the amplitude of the wave varies gradually with time and space, as is shown in Fig. 3.1. This is a phenomenon of wave amplitude modulation caused by modulational instability, which will be discussed in Chap. 9.

If the wave amplitude varies periodically in this manner, the envelope of the wave also becomes a periodic function. This indicates that the amplitude h_0 of (2.5) is a function of the space x and time t. Therefore, the wave height $h(x,t)$ can be written as

$$h(x,t) = h_0(x,t)\ \cos(kx - \omega t) \quad . \tag{3.1}$$

Here, $h_0(x,t)$ is a function describing the wave envelope. By the definition of an envelope, $h_0(x,t)$ is a function which varies much more slowly in time and space as compared to the phase of the wave $kx - \omega t$.

In this section we consider the mathematical description of this envelope function. In particular, we consider the envelope function of a light wave in dielectric media.

The refractive index n of a medium is given by the ratio of the speed of light c in the vacuum to the phase velocity of the wave in the medium,

$$\frac{ck}{\omega} = n \quad . \tag{3.2}$$

The refractive index of a dielectric medium such as a glass fiber changes as a function of the frequency of the light wave because of its molecular structure. A medium where the refractive index varies as a function of the frequency is called a dispersive medium. Because of the dispersive property of the fiber, the envelope of a modulated light wave, such as a light pulse, is distorted owing to the fact that each frequency component of the spectrum has a different velocity.

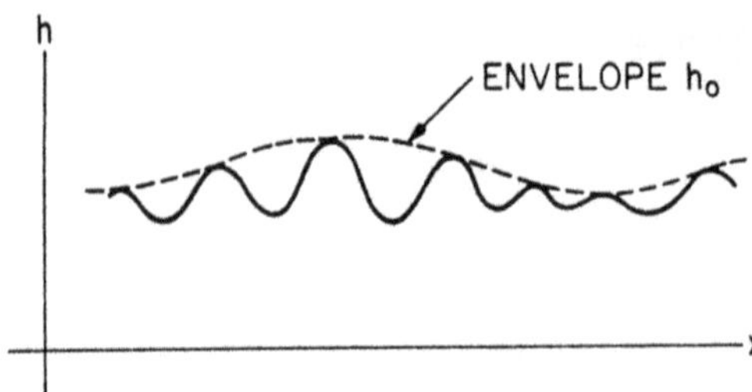

Fig. 3.1. Envelope of a water surface

Let us now write the electric field of the light wave with slowly varying amplitude in terms of its complex amplitude $E(x,t)$,

$$E = \mathrm{Re}\{E(x,t)\ \exp[\mathrm{i}\,(k_0 x - \omega_0 t)]\} \quad . \tag{3.3}$$

Here, Re denotes the real part and k_0 and ω_0 represent the wave number and angular frequency of the carrier wave. The fact that the envelope function $E(x,t)$ is a slowly varying function of time and space indicates that the frequency spectrum of the electric field E has a localized structure around the carrier frequency ω_0, as shown in Fig. 3.2. In this figure, $\Delta\omega_0$ shows the width of the frequency spectrum of the envelope function E.

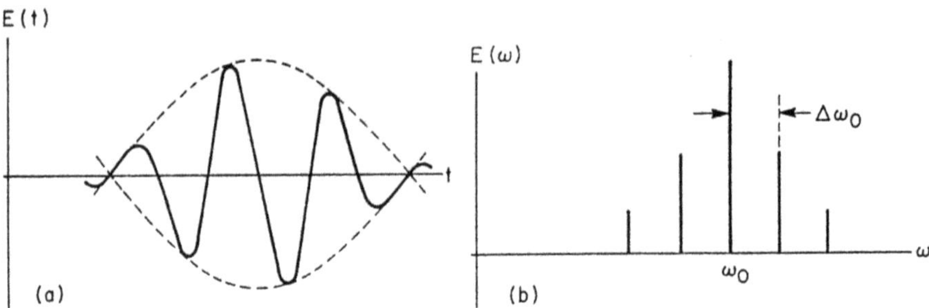

Fig. 3.2. Modulated wave (a) and its frequency spectrum (b)

3.2 Wave Equation for the Envelope Function

If dispersion exists in a fiber, the refractive index n becomes a function of the wave frequency of the light wave. In this case, quantities such as $\partial n/\partial\omega$, $\partial^2 n/\partial\omega^2$ ... have a finite value. As a result, derivatives of the wave number k with respect to the frequency ω, $\partial k/\partial\omega$, $\partial^2 k/\partial\omega^2$ ... also have finite values. If we make use of this fact, we can find the propagation property of the side band (which is the Fourier component around the carrier frequency ω_0 produced by the modulation).

Let us expand the wave number $k(= n\omega/c)$ around the carrier frequency ω_0, utilizing (3.2) and the fact that the refractive index n is a function of ω,

$$k - k_0 = \left.\frac{\partial k}{\partial\omega}\right|_{\omega_0}(\omega-\omega_0) + \frac{1}{2}\left.\frac{\partial^2 k}{\partial\omega^2}\right|_{\omega_0}(\omega-\omega_0)^2 + \frac{1}{6}\left.\frac{\partial^3 k}{\partial\omega^3}\right|_{\omega_0}(\omega-\omega_0)^3 + \dots \ . \tag{3.4}$$

This equation describes the wave number of the frequency component of the modulated wave whose frequency deviates slightly from the carrier frequency ω_0.

Given the fact that the wave envelope function $E(x,t)$ is a slowly varying function in x and t, we Fourier transform this function using the Fourier space variables $\Delta\omega\,(=\omega-\omega_0)$. This represents a small frequency shift of the side band from the carrier frequency and $\Delta k\ (=k-k_0)$, which represents the shift of the wave number

$$\overline{E}(\Delta k,\Delta\omega)=\int_{-\infty}^{\infty}\int_{-\infty}^{\infty}E(x,t)\,\mathrm{e}^{\mathrm{i}(\Delta\omega t-\Delta kx)}\,dx\,dt \quad . \tag{3.5}$$

The inverse transform is given by

$$E(x,t)=\frac{1}{(2\pi)^2}\int_{-\infty}^{\infty}\int_{-\infty}^{\infty}\overline{E}(\Delta k,\Delta\omega)\,\mathrm{e}^{-\mathrm{i}(\Delta\omega t-\Delta kx)}\,d(\Delta k)\,d(\Delta\omega) \quad . \tag{3.6}$$

From (3.5) and (3.6), we see that $\partial E/\partial t$ and $\partial E/\partial x$ can be identified as the Fourier transform of $-\mathrm{i}\Delta\omega\overline{E}$ and $\mathrm{i}\Delta k\overline{E}$ respectively. Using this fact, we rewrite $\Delta\omega$ and Δk as $\mathrm{i}\partial/\partial t$ and $-\mathrm{i}\partial/\partial x$, respectively, and express (3.4) in terms of these operators,

$$-\mathrm{i}\frac{\partial}{\partial x}=\mathrm{i}k'\frac{\partial}{\partial t}-\frac{k''}{2}\frac{\partial^2}{\partial t^2}-\mathrm{i}\frac{k'''}{6}\frac{\partial^3}{\partial t^3}+\dots \quad . \tag{3.4$'$}$$

If we operate (3.4$'$) on the envelope function $E(x,t)$ and keep terms up to the order of the second derivative with respect to the wave number, we then have

$$\mathrm{i}\left(\frac{\partial}{\partial x}+k'\frac{\partial}{\partial t}\right)E-\frac{k''}{2}\frac{\partial^2 E}{\partial t^2}=0 \quad . \tag{3.7}$$

Here, $k'=\partial k/\partial\omega|_{\omega_0}$, $k''=\partial^2 k/\partial\omega^2|_{\omega_0}$. The group velocity v_g of a modulated wave is given by

$$v_\mathrm{g}=\frac{\partial\omega}{\partial k}=\frac{1}{k'} \tag{3.8}$$

and hence, k'' in (3.7) is given by

$$k''=\frac{\partial}{\partial\omega}\left(\frac{1}{v_\mathrm{g}}\right)=-\frac{1}{v_\mathrm{g}^2}\frac{\partial v_\mathrm{g}}{\partial\omega} \quad . \tag{3.9}$$

This expression indicates that k'' is given by the frequency dependence of the wave group velocity. Therefore, k'' represents the dispersion property of the group velocity of the wave.

If $k''=0$, the solution of (3.7) can be expressed in terms of an arbitrary function of $x-t/k'=x-v_\mathrm{g}t$, $E(x-v_\mathrm{g}t)$. This represents the fact that the envelope of the light wave propagates at group speed. Proceeding from this fact, let us use a new coordinate system which moves at group velocity,

$$\xi=\varepsilon^2 x \quad , \qquad \tau=\varepsilon(t-k'x) \quad . \tag{3.10}$$

Equation (3.7) then becomes

$$\mathrm{i}\frac{\partial E}{\partial \xi} - \frac{k''}{2}\frac{\partial^2 E}{\partial \tau^2} = 0 \quad . \tag{3.11}$$

Here, ε is a small quantity $\Delta\omega_0/\omega_0$ which designates the relative width of the spectrum. Equation (3.11) shows that the envelope function of the wave deforms in proportion to the distance of propagation, due to group velocity dispersion.

If a short optical pulse is transmitted through an optical fiber, $\partial^2 E/\partial\tau^2$ becomes larger in inverse proportion to the square of the pulse width, and the pulse shape distorts during the transmission. In the case of a fiber with a cross-sectional dimension which is large as compared to the wavelength of the light, the group velocity dispersion k'' is decided by the property of the glass material itself and, at a wavelength approximately equal to 1.3 μm, k'' becomes 0. At shorter wavelengths, k'' becomes positive, i.e., the group velocity is a decreasing function of the frequency, while at longer wavelengths, k'' becomes negative and the group velocity is an increasing function of the frequency. The region of the wavelength corresponding to positive k'' is called a normal dispersion region, while that corresponding to negative k'' is called a negative, or anomalous dispersion region.

It is known that the fiber loss produced by Rayleigh scattering and molecular vibration becomes minimum at a wavelength approximately equal to 1.5 μm. Therefore, if one chooses the wavelength that corresponds to the minimum loss of the fiber, the group dispersion k'' becomes negative and the pulse shape distorts due to the group velocity dispersion.

As will be shown in Sect. 3.4, group velocity dispersion depends not only on the property of the glass material, but also on the wave guide property of the fiber. Therefore, the group dispersion property can be modified by an appropriate choice of the variation of the refractive index across the fiber cross-section. By using this method, a dispersion shifted fiber in which the group velocity dispersion vanishes at a wavelength of 1.5 μm (where the loss rate becomes minimum) can be developed. Technically, the group dispersion property of a fiber is expressed by a constant D in units of ps/(nm km). k'' can be expressed in terms of this quantity D,

$$k'' = D\frac{\lambda^2}{2\pi c} \quad . \tag{3.12}$$

Here, λ is the wavelength of the light in a vacuum, and c the speed of light. If we write the wavelength at which the group velocity dispersion vanishes as λ_0, D has a value of approximately ∓ 10 ps/(nm km) at $\lambda = \lambda_0 \pm 0.1$ μm. By using the observed data for SiO_2, k'' can also be expressed approximately as [3.1]

$$k'' = -5.3 \times 10^{-2}\left(1 - \frac{\lambda_0}{\lambda}\right)\frac{\lambda}{2\pi c^2} \quad . \tag{3.12$'$}$$

If we take a nominal value of $\lambda \simeq 1.5$ μm, $D = -10$ ps/(nm km), the value k'' becomes approximately -10 ps^2/km. This indicates that an optical pulse in a fiber with a pulse width of a few picoseconds would distort if it propagated over a distance of 1 km.

3.3 Nonlinear Schrödinger Equation and a Solitary Wave Solution

Let us now consider the nonlinear effect of a dielectric material in light wave propagation. In an ordinary dielectric material, the refractive index increases in proportion to the square of the electric field. This property is known as Kerr effect [3.2]. If we consider the Kerr effect, the refractive index can be written as

$$n = n_0(\omega) + n_2 \left|E\right|^2 \quad .$$

Here, the magnitude of E represents the effective amplitude of the electric field, and n_2 is called the Kerr coefficient. For a glass fiber, n_2 is known to have a value of approximately $1.2 \times 10^{-22}\ \mathrm{m^2/V^2}$. The Kerr effect originates from a deformation of the electron orbits in glass molecules due to the applied electric field. As a result, it has an extremely short response time of the order of 10^{-15} seconds.

Since a standard optical fiber has an effective cross-sectional area of approximately $60\ \mu\mathrm{m}^2$, an optical power of approximately 100 mW would produce an electric field of the light of the order of 10^6 V/m. Consequently, the refractive index increases by a factor of approximately 10^{-10}. Because of this, the wave number of the light wave in a fiber changes by a factor of $n_2 \left|E\right|^2 \omega/c = 2\pi n_2/\lambda \left|E\right|^2$, as indicated by (3.2). If we take a nominal wavelength of 1.5×10^{-6}m, the change in the wave number also becomes of the order of km^{-1}. If we incorporate this nonlinear effect into the right-hand side of (3.4), (3.11) is now modified to

$$\mathrm{i}\,\frac{\partial E}{\partial \xi} - \frac{k''}{2}\,\frac{\partial^2 E}{\partial \tau^2} + g\,\frac{\left|E\right|^2 E}{\varepsilon^2} = 0 \quad . \tag{3.13}$$

In this expression, $g = 2\pi n_2 \alpha/\lambda$ and α represent the reduction factor due to the fact that the light intensity varies in the cross-section of the fiber and takes a value of approximately 1/2 in most cases. The exact expression for α will be derived in Sect. 3.4.

If we compare (3.13) with the well-known Schrödinger equation,

$$\mathrm{i}\,\frac{\partial \psi}{\partial t} + \frac{\partial^2 \psi}{\partial x^2} + V\psi = 0 \quad , \tag{3.14}$$

and replace V by $\left|E\right|^2$, t by ξ and x by τ, we recognize the exact similarity. The fact that the potential V of the Schrödinger equation is represented by $\left|E\right|^2$ originates from the fact that the refractive index varies in proportion to the magnitude E^2. For this reason, (3.14) is called the nonlinear Schrödinger equation.

Since, in the Schrödinger equation, V expresses the potential which traps the quasi-particle represented by the wave function ψ, the fact that V increases in proportion to $|E|^2$ indicates that if k'' is negative, the depth of the trapping potential increases in proportion to the intensity of the light. Therefore, if k'' is negative, the potential which is proportional to $|E|^2$ has the effect of trapping the wave energy, which otherwise tends to spread due to the dispersion. We call this the self-trapping of a wave. When the wave intensity is concentrated in a local region due to this self-trapping, conservation of energy requires that the pulse width becomes shorter in inverse proportion to $|E|^2$. As this trapping progresses the dispersion effect, which increases in inverse proportion to the square of the pulse width, becomes more effective and the effect of the expansion of the pulse out-weighs the contraction effect due to the nonlinearity, causing the pulse to spread. At some values of the pulse width, the spreading effect due to dispersion and the self-trapping effect due to nonlinearity balance, and a stationary pulse can be formed.

In order to derive this stationary solution, let us first normalize (3.13) in the following form:

$$\mathrm{i}\,\frac{\partial q}{\partial Z}+\frac{1}{2}\frac{\partial^2 q}{\partial T^2}+|q|^2\,q=0 \quad . \tag{3.15}$$

Here

$$q=\frac{\sqrt{g}\,\lambda}{\varepsilon}\,E \tag{3.16}$$

$$T=\frac{\tau}{(-\lambda k'')^{1/2}} \tag{3.17}$$

$$Z=\frac{\xi}{\lambda} \quad . \tag{3.18}$$

If we use earlier examples such as $D=-10\mathrm{ps}/(\mathrm{nm\,km})$, $\lambda=1.5\,\mu\mathrm{m}$, the choice of $10^{-4.5}$ for the normalized amplitude ε in these expressions gives relations such that $Z=1$ corresponds to $\xi=1.5\,\mathrm{km}$, $q=1$ corresponds to $E=2\times10^6\,\mathrm{V/m}$, and $T=1$ corresponds to $t=2\,\mathrm{ps}$.

We look for a localized solution of $|q|$ which is stationary in Z, i.e., a stationary shape of the packet. Since we are interested in a localized solution, we ensure that the solution will be single-humped by imposing the following conditions:

1) $|q|^2$ is bounded by the two limits ϱ_s and ϱ_D;
2) at $|q|^2=\varrho_s$, $|q|^2$ is an extremum, i.e., at $|q|^2=\varrho_s$, $\partial|q|^2/\partial T=0$, but $\partial^2|q|^2/\partial T^2\neq 0$;
3) ϱ_D is the asymptotic value of $|q|^2$ as $T\to\pm\infty$, i.e., at $|q|^2=\varrho_D$, $\partial^n|q|^2/\partial T^n=0$, $n=1,2,\dots$.

We now look for a solution of (3.15) which satisfies these conditions. In order to achieve this, we introduce two real variables, ϱ and σ, which represent the real and imaginary parts of q:

$$q(T,Z)=\sqrt{\varrho(T,Z)}\,\mathrm{e}^{\mathrm{i}\sigma(T,Z)} \quad . \tag{3.19}$$

Substituting this expression into (3.15), we have

$$\frac{\partial \varrho}{\partial Z} + \frac{\partial}{\partial T}\left(\varrho \frac{\partial \sigma}{\partial T}\right) = 0 \tag{3.20}$$

and

$$\frac{1}{8}\frac{d}{d\varrho}\left[4\varrho^2 + \frac{1}{\varrho}\left(\frac{\partial \varrho}{\partial T}\right)^2\right] = \frac{\partial \sigma}{\partial Z} + \frac{1}{2}\left(\frac{\partial \sigma}{\partial T}\right)^2 \quad . \tag{3.21}$$

The stationary condition for $|q|^2\,(=\varrho)$ gives $\partial\varrho/\partial Z = 0$. Hence, from (3.20), we have

$$\varrho \frac{\partial \sigma}{\partial T} = c(Z) \quad . \tag{3.22}$$

We now show that the only choice possible for the integration constant $c(Z)$ is a constant independent of Z. In order to prove this, we note that the left-hand side of (3.21) is a function of T alone, and thus

$$\frac{\partial \sigma}{\partial Z} + \frac{1}{2}\left(\frac{\partial \sigma}{\partial T}\right)^2 = f(T) \quad .$$

By taking derivations with respect to Z and T, we have

$$\frac{\partial^3 \sigma}{\partial Z^2\,\partial T} - \frac{1}{\varrho^3}\frac{d\varrho}{dT}\frac{dc^2}{dZ} = 0 \quad ,$$

while from (3.22),

$$\frac{\partial^3 \sigma}{\partial Z^2\,\partial T} = \frac{1}{\varrho}\frac{d^2 c}{dZ^2} \quad .$$

Hence,

$$\frac{1}{\varrho}\frac{d^2 c}{dZ^2} - \frac{1}{\varrho^3}\frac{d\varrho}{dT}\frac{dc^2}{dZ} = 0 \quad ,$$

or

$$\frac{d^2 c}{dZ^2}\Big/\frac{dc^2}{dZ} = \frac{1}{\varrho^2}\frac{d\varrho}{dT} = \text{const} \quad .$$

Since we cannot accept the solution $\varrho^{-2}\,d\varrho/dT = \text{const}$, the only alternative choice is $c(Z) = \text{const}$. Consequently, (3.22) becomes

$$\varrho \frac{\partial \sigma}{\partial T} = c_1\,(\text{const}) \quad , \tag{3.23}$$

or

$$\sigma = \int \frac{c_1}{\varrho} \, dT + A(Z) \quad . \tag{3.24}$$

Because $\partial\sigma/\partial T$ has been proved to be a function of T alone, from (3.21), $\partial\sigma/\partial Z$ should also be a function of T. Hence, we take dA/dZ to be constant $(= \Omega)$,

$$\sigma = \int \frac{c_1}{\varrho} \, dT + \Omega Z \quad . \tag{3.25}$$

If we use this expression in (3.21), we obtain the following ordinary differential equation for $\varrho(T)$:

$$\left(\frac{d\varrho}{dT}\right)^2 = -4\varrho^3 + 8\Omega\varrho^2 + c_2\varrho - 4c_1^2 \quad . \tag{3.26}$$

We now seek a solution of this equation, subject to the conditions 1 to 3.

In order to satisfy condition 1, $d\varrho/dT$ should vanish only at two values of ϱ, ϱ_{D} and ϱ_{s}. In addition, for the root at ϱ_{D} to represent an asymptotic value of ϱ, it should be a double root. These conditions are met only for $-4c_1^2 \geq 0$, or $c_1 = 0$ and hence, also $c_2 = 0$. Equation (3.26) then reduces to

$$\left(\frac{d\varrho}{dT}\right)^2 = -4\varrho^3 + 8\Omega\varrho^2 = -4\varrho^2(\varrho - \varrho_{\mathrm{s}}) \quad , \tag{3.27}$$

where

$$\varrho_{\mathrm{s}} = 2\Omega \quad . \tag{3.28}$$

Equation (3.27) can then easily be integrated to give

$$\varrho = \varrho_0 \operatorname{sech}^2(\sqrt{\varrho_0}\, T) \quad , \tag{3.29}$$

where $\varrho_0 (= \varrho_{\mathrm{s}}) = 2\Omega$, $\Omega > 0$ and

$$\sigma = \Omega Z = \varrho_0/2 \quad . \tag{3.30}$$

The Schrödinger equation (3.15) can now be shown to be satisfied by another function $q'(T, Z)$, given by

$$q'(T, Z) = \exp[-\mathrm{i}(\kappa T + \tfrac{1}{2}\kappa^2 Z)]\, \phi(T + \kappa Z, Z) \quad .$$

With the additional independent variable κ, and phase constants θ_0 and σ_0, the solitary envelope solution becomes

$$q(T, Z) = \eta \operatorname{sech} \eta(T + \kappa Z - \theta_0) \exp\left\{-\mathrm{i}\kappa T + \frac{\mathrm{i}}{2}(\eta^2 - \kappa^2)Z - \mathrm{i}\sigma_0\right\} \ , \tag{3.31}$$

where $\sqrt{\varrho_0}$ is replaced by η.

The solitary wave solution represented by (3.31) has four parameters. They are η which represents the amplitude and pulse width of the solitary wave, κ

which represents its speed (we should note that this speed represents a deviation from the group velocity), and 2 parameters which represent the phase constants θ_0 and σ_0. We note here that the pulse height η is inversely proportional to the pulse width η^{-1}, and that the constant κ, which represents the speed of the pulse transmission, is independent of the pulse height η. In this particular respect, the latter fact differs from the KdV soliton where the speed of the soliton is proportional to the pulse height. *Zakharov* and *Shabat* [3.3] succeeded in solving the nonlinear Schrödinger equation by considering the inverse scattering problem, and thereby demonstrated that the solution of the equation can be described by the combination of solitary wave solutions expressed in (3.31), and a continuous wave. In this respect (3.31) is the soliton solution of the nonlinear Schrödinger equation. For this reason, the solitary wave solution expressed in (3.31) is called the envelope soliton.

The envelope soliton exists when the wave length is in the region of anomalous dispersion, where k'' is negative. On the contrary, in the region of normal dispersion where k'' is positive, the portion without a continuous light wave, i.e., the place where light is absent, becomes a soliton [3.4]. For this reason, the soliton solution in the range of positive k'' is called a dark soliton. We will discuss the dark soliton in Chap. 10.

3.4 Derivation of the Envelope Equation for a Light Wave in a Fiber

For further background information on the material discussed in this section see [3.5, 6]. Now we reduce the Maxwell equations (3-dimensional vector equations) to the nonlinear Schrödinger equation by introducing an appropriate scale of coordinates based on the physical setting of a cylindrical dielectric guide. The higher order terms describe the linear and nonlinear dispersion, as well as dissipation effects (1-dimensional scalar equation). The method used here is based on the asymptotic perturbation technique developed by *Taniuti* et al. [3.7] (the so-called reductive perturbation method), and gives a consistent scheme for the derivation of the nonlinear Schrödinger equation and the higher order terms.

The electric field $\boldsymbol{E}$, with the dielectric constant $\varepsilon_0 \overleftrightarrow{\chi}$, satisfies the Maxwell equation,

$$\nabla \times \nabla \times \boldsymbol{E} = -\frac{1}{c^2}\frac{\partial^2}{\partial t^2}\boldsymbol{D} \quad . \tag{3.32}$$

Here, the displacement vector $\boldsymbol{D} = \overleftrightarrow{\chi} * \boldsymbol{E}$ is the Fourier transform of $\hat{\boldsymbol{D}}(\omega)$ defined through (3.33) and is given by the convolution integral of $\overleftrightarrow{\chi}(t)$ and $\boldsymbol{E}(t)$

$$\begin{aligned}\overleftrightarrow{\chi} * \boldsymbol{E}(t) = &\int_{-\infty}^{t} dt_1\, \overleftrightarrow{\chi}^{(0)}(t-t_1)\,\boldsymbol{E}(t_1) + \int_{-\infty}^{t} dt_1 \int_{-\infty}^{t} dt_2 \int_{-\infty}^{t} dt_3 \\ &\times \overleftrightarrow{\chi}^{(2)}(t-t_1\,, t-t_2\,, t-t_3)\{\boldsymbol{E}(t_1)\cdot\boldsymbol{E}(t_2)\}\,\boldsymbol{E}(t_3) \\ &+ (\text{higher nonlinear terms}) \quad .\end{aligned} \tag{3.33}$$

Here, the second term describing the nonlinear polarization includes the Kerr and Raman effects with proper retardation. The dielectric tensors $\overleftrightarrow{\chi}^{(0)}$, $\overleftrightarrow{\chi}^{(2)}$ are dependent on the spatial coordinates in the transverse direction of the fiber axis. We write (3.32) in the following form:

$$\nabla^2 \boldsymbol{E} - \frac{1}{c^2}\frac{\partial^2}{\partial t^2}\boldsymbol{D} = \nabla(\nabla \cdot \boldsymbol{E}) \quad . \tag{3.34}$$

It should be noted that $\nabla \cdot \boldsymbol{E}$ in (3.34) is not zero, since $\nabla \cdot \boldsymbol{D} = 0$ (the constraint for $\boldsymbol{D}$ in Maxwell's equation) implies that $\overleftrightarrow{\chi} * (\nabla \cdot \boldsymbol{E}) = -(\nabla \overleftrightarrow{\chi} *) \cdot \boldsymbol{E} \neq 0$, namely, the electric field cannot be described by either the TE or TM modes.

Since our purpose is to reduce (3.34) in the sense of an asymptotic perturbation, it is convenient to write it in the following matrix form,

$$L\tilde{\boldsymbol{E}} = 0 \quad , \tag{3.35}$$

where $\tilde{\boldsymbol{E}}$ represents a column vector, i.e. $(E_r, E_\theta, E_z)^t$. In cylindrical coordinates, where the z axis is the axial direction of the fiber, the matrix L consisting of the three parts $L = L_\mathrm{a} + L_\mathrm{b} - L_\mathrm{c}$ is defined by

$$L_\mathrm{a} = \begin{pmatrix} \nabla_\perp^2 - \frac{1}{r^2} & -\frac{2}{r^2}\frac{\partial}{\partial\theta} & 0 \\ \frac{2}{r^2}\frac{\partial}{\partial\theta} & \nabla_\perp^2 - \frac{1}{r^2} & 0 \\ 0 & 0 & \nabla_\perp^2 \end{pmatrix} \quad , \tag{3.36a}$$

$$L_\mathrm{b} = \left(\frac{\partial^2}{\partial z^2} - \frac{1}{c^2}\frac{\partial^2}{\partial t^2}\overleftrightarrow{\chi} *\right)\begin{pmatrix} 1 & 0 & 0 \\ 0 & 1 & 0 \\ 0 & 0 & 1 \end{pmatrix} \quad , \tag{3.36b}$$

$$L_\mathrm{c} = \begin{pmatrix} \frac{\partial}{\partial r}\frac{1}{r}\frac{\partial}{\partial r}r & \frac{\partial}{\partial r}\frac{1}{r}\frac{\partial}{\partial\theta} & \frac{\partial^2}{\partial r\,\partial z} \\ \frac{1}{r^2}\frac{\partial^2}{\partial r\,\partial\theta}r & \frac{1}{r^2}\frac{\partial^2}{\partial\theta^2} & \frac{1}{r}\frac{\partial^2}{\partial\theta\,\partial z} \\ \frac{1}{r}\frac{\partial^2}{\partial r\,\partial z}r & \frac{1}{r}\frac{\partial^2}{\partial\theta\,\partial z} & \frac{\partial^2}{\partial z^2} \end{pmatrix} \quad . \tag{3.36c}$$

It should be noted that these matrices imply that

$$L_\mathrm{a}\tilde{\boldsymbol{E}} = \nabla_\perp^2 \boldsymbol{E} \equiv \left(\frac{1}{r}\frac{\partial}{\partial r}r\frac{\partial}{\partial r} + \frac{1}{r^2}\frac{\partial^2}{\partial\theta^2}\right)\boldsymbol{E} \quad ,$$

$$L_\mathrm{b}\tilde{\boldsymbol{E}} = \left(\frac{\partial^2}{\partial z^2} - \frac{1}{c^2}\frac{\partial^2}{\partial t^2}\overleftrightarrow{\chi} *\right)\boldsymbol{E}$$

and

$$L_\mathrm{c}\tilde{\boldsymbol{E}} = \nabla(\nabla \cdot \boldsymbol{E}) \quad .$$

We consider the electric field as an almost monochromatic wave propagating along the z axis with wave number k_0 and angular frequency ω_0, i.e., the field $\tilde{\boldsymbol{E}}$ is assumed to be in the expansion form,

$$\tilde{\boldsymbol{E}}(r,\theta,z,t)=\sum_{l=-\infty}^{\infty}\tilde{\boldsymbol{E}}_l(r,\theta,\xi,\tau,\varepsilon)\,\exp\{\mathrm{i}(k_l z-\omega_l t)\} \tag{3.37}$$

with $\tilde{\boldsymbol{E}}_{-l}=\tilde{\boldsymbol{E}}_l^*$ (complex conjugate). $k_l=l\,k_0$, $\omega_l=l\,\omega_0$ and the summation is taken over all the harmonics generated by the nonlinear response of the polarization, $\tilde{\boldsymbol{E}}_l(r,\theta,\xi,\tau,\varepsilon)$ being the envelope of the l-th harmonic which changes slowly in z and t. Here, the slow variables ξ and τ are defined by

$$\xi=\varepsilon^2 z \quad , \qquad \tau=\varepsilon\Big(t-\frac{z}{v_g}\Big) \tag{3.10$'$}$$

where v_g is the group velocity of the wave given below. We note that $\tilde{\boldsymbol{E}}_l$ in expression (3.37) represents twice the amplitude of the electric field defined in (3.3). Since the radius of the fiber has the same order as the wavelength $(2\pi/k_0)$, the scale for the transverse coordinates (r,θ) is of order 1. In this scale of the coordinates of (3.10$'$), the behaviour of the field can be deduced from the balance between the nonlinearity and dispersion which results in the formation of optical solitons confined in the transverse direction.

If we proceed from (3.37) and (3.10$'$), we find that the displacement, $\boldsymbol{D}=\overset{\leftrightarrow}{\chi}*\boldsymbol{E}=\sum\boldsymbol{D}_l\,\exp\{\mathrm{i}(k_l z-\omega_l t)\}$, is given by

$$\begin{aligned}\tilde{\boldsymbol{D}}_l=&\chi_l^{(0)}\,\tilde{\boldsymbol{E}}_l+\varepsilon\mathrm{i}\,\frac{\partial\chi_l^{(0)}}{\partial\omega_l}\,\frac{\partial\tilde{\boldsymbol{E}}_l}{\partial\tau}-\varepsilon^2\,\frac{1}{2}\,\frac{\partial^2\chi_l^{(0)}}{\partial\omega_l^2}\,\frac{\partial^2\tilde{\boldsymbol{E}}_l}{\partial\tau^2}\\&-\varepsilon^3\,\frac{\mathrm{i}}{6}\,\frac{\partial^3\chi_l^{(0)}}{\partial\omega_l^3}\,\frac{\partial^3\tilde{\boldsymbol{E}}_l}{\partial\tau^3}+\sum_{l_1+l_2+l_3=l}\Big\{\chi^{(2)}_{l_1l_2l_3}\,(\tilde{\boldsymbol{E}}_{l_1}\cdot\tilde{\boldsymbol{E}}_{l_2})\,\tilde{\boldsymbol{E}}_{l_3}\\&+\mathrm{i}\sum_{i=1}^{3}\Big(\frac{\partial}{\partial\omega_{l_1}}\,\chi^{(2)}_{l_1l_2l_3}\Big)\,\frac{\partial}{\partial\tau_i}\,(\tilde{\boldsymbol{E}}_{l_1}\cdot\tilde{\boldsymbol{E}}_{l_2})\,\tilde{\boldsymbol{E}}_{l_3}\Big\}+\ldots\end{aligned} \tag{3.38}$$

where $\chi_l^{(0)}$ is the Fourier coefficient $\hat{\chi}^{(0)}(\Omega)$ of $\chi^{(0)}(t)$ at $\Omega=\omega_l$, i.e., $\chi_l^{(0)}=\hat{\chi}^{(0)}(\omega_l)$, and $\chi^{(2)}_{l_1l_2l_3}$ is the Fourier coefficient $\hat{\chi}^{(2)}(\Omega_1,\Omega_2,\Omega_3)$ of $\chi^{(2)}(t_1,t_2,t_3)$ at $\Omega_1=\omega_{l_1}$, $\Omega_2=\omega_{l_2}$, $\Omega_3=\omega_{l_3}$ and $\partial(\tilde{\boldsymbol{E}}_{l_1}\cdot\tilde{\boldsymbol{E}}_{l_2})\,\tilde{\boldsymbol{E}}_{l_3}/\partial\tau_1=(\partial\tilde{\boldsymbol{E}}_{l_1}/\partial\tau\cdot\tilde{\boldsymbol{E}}_{l_2})\,\tilde{\boldsymbol{E}}_{l_3}$, $\partial(\tilde{\boldsymbol{E}}_{l_1}\cdot\tilde{\boldsymbol{E}}_{l_2})\,\tilde{\boldsymbol{E}}_{l_3}/\partial\tau_2=(\tilde{\boldsymbol{E}}_{l_1}\cdot\partial\tilde{\boldsymbol{E}}_{l_2}/\partial\tau)\,\tilde{\boldsymbol{E}}_{l_3}$ and so on. The last term in (3.38) represents the retarded response of the nonlinear polarization which gives both the higher order nonlinear dispersion and dissipation. We note that owing to the dispersion properties of the dielectric constant, the real space wave equation contains terms with higher-order time derivatives. From (3.10$'$) and (3.38), (3.35) can be written in the following expansion form:

$$
\begin{aligned}
&L_l \tilde{\boldsymbol{E}}_l + \mathrm{i}\varepsilon \frac{\partial L_l}{\partial \omega_l} \frac{\partial \tilde{\boldsymbol{E}}_l}{\partial \tau} - \frac{\varepsilon^2}{2} \frac{\partial^2 L_l}{\partial \omega_l^2} \frac{\partial^2 \tilde{\boldsymbol{E}}_l}{\partial \tau^2} - \frac{\mathrm{i}}{6} \varepsilon^3 \frac{\partial^3 L_l}{\partial \omega_l^3} \frac{\partial^3 \tilde{\boldsymbol{E}}_l}{\partial \tau^3} \\
&\quad - \varepsilon^2 \left\{ \frac{\partial}{\partial k_l} \left(L_l - \frac{\omega_l^2 \chi_l^{(0)}}{c^2} \right) \right\} \\
&\quad \times \left(\mathrm{i} \frac{\partial \tilde{\boldsymbol{E}}_l}{\partial \xi} - \frac{1}{2} \frac{\partial^2 k_l}{\partial \omega_l^2} \frac{\partial^2 \tilde{\boldsymbol{E}}_l}{\partial \tau^2} - \varepsilon \frac{\mathrm{i}}{6} \frac{\partial^3 k_l}{\partial \omega_l^3} \frac{\partial^3 \tilde{\boldsymbol{E}}_l}{\partial \tau^3} \right) \\
&\quad - \varepsilon^3 \left\{ \frac{\partial^2}{\partial \omega_l \partial k_l} \left(L_l - \frac{\omega_l^2 \chi_l^{(0)}}{c^2} \right) \right\} \mathrm{i} \frac{\partial}{\partial \tau} \left(\mathrm{i} \frac{\partial \tilde{\boldsymbol{E}}_l}{\partial \xi} - \frac{1}{2} \frac{\partial^2 k_l}{\partial \omega_l^2} \frac{\partial^2 \tilde{\boldsymbol{E}}_l}{\partial \tau^2} \right) \\
&\quad + \frac{1}{c^2} \sum_{l_1 + l_2 + l_3 = l} \Bigg[\omega_l^2 \chi_{l_1 l_2 l_3}^{(2)} (\tilde{\boldsymbol{E}}_{l_1} \cdot \tilde{\boldsymbol{E}}_{l_2}) \tilde{\boldsymbol{E}}_{l_3} \qquad\qquad (3.39) \\
&\quad + \mathrm{i}\varepsilon \sum_{i=1}^{3} \left\{ \frac{\partial}{\partial \omega_{l_i}} (\omega_l^2 \chi_{l_1 l_2 l_3}^{(2)}) \right\} \frac{\partial}{\partial \tau_i} (\tilde{\boldsymbol{E}}_{l_1} \cdot \tilde{\boldsymbol{E}}_{l_2}) \tilde{\boldsymbol{E}}_{l_3} \Bigg] \\
&\quad + \mathrm{i}\varepsilon \left(\frac{1}{v_g} - \frac{\partial k_l}{\partial \omega_l} \right) \left\{ \frac{\partial}{\partial k_l} \left(L_l - \frac{\omega_l^2 \chi_l^{(0)}}{c^2} \right) \right\} \frac{\partial \tilde{\boldsymbol{E}}_l}{\partial \tau} \\
&\quad + \varepsilon^2 \left(\frac{1}{v_g} - \frac{\partial k_l}{\partial \omega_l} \right) \begin{bmatrix} 1 & 0 & 0 \\ 0 & 1 & 0 \\ 0 & 0 & 0 \end{bmatrix} \frac{\partial}{\partial \tau} \left\{ \left(\frac{1}{v_g} + \frac{\partial k_l}{\partial \omega_l} \right) \frac{\partial \tilde{\boldsymbol{E}}_l}{\partial \tau} - 2\varepsilon \frac{\partial \tilde{\boldsymbol{E}}_l}{\partial \xi} \right\} \\
&\quad + \ldots = 0 \quad ,
\end{aligned}
$$

where L_l is L in (3.36) with the replacement $\partial/\partial z = \mathrm{i}k_l$, $\partial/\partial t = -\mathrm{i}\omega_l$ and $\chi * = \chi_l^{(0)}$. It should be noted that the operator L_l is self-adjoint in the sense of the following inner product:

$$
(\tilde{\boldsymbol{U}}, L\tilde{\boldsymbol{V}}) = \int \tilde{\boldsymbol{U}}^* \cdot L\tilde{\boldsymbol{V}}\, dS = \int L^* \tilde{\boldsymbol{U}}^* \cdot \tilde{\boldsymbol{V}}\, dS = (L\tilde{\boldsymbol{U}}, \tilde{\boldsymbol{V}}) \quad , \qquad (3.40)
$$

where $dS = r\, dr\, d\theta$ and $\tilde{\boldsymbol{U}}, \tilde{\boldsymbol{V}} \to 0$ as $\| x_\perp \| \to \infty$.

We now assume that $\tilde{\boldsymbol{E}}_l(r, \theta, \xi, \tau; \varepsilon)$ can be expanded in terms of ε,

$$
\tilde{\boldsymbol{E}}_l(r, \theta, \xi, \tau; \varepsilon) = \sum_{n=1}^{\infty} \varepsilon^n \tilde{\boldsymbol{E}}_l^{(n)}(r, \theta, \xi, \tau) \quad . \qquad (3.41)
$$

Then, from (3.35), (3.37), (3.10′) and (3.38) we have, at order ε,

$$
L_l \tilde{\boldsymbol{E}}_l^{(1)} = 0 \quad . \qquad (3.42)
$$

In (3.42) we consider a mono-mode fiber in which there is only one bound state with eigenvalue k_0^2 (i.e. $l = \pm 1$) and eigenfunction $\tilde{\boldsymbol{U}} = \tilde{\boldsymbol{U}}(r, \theta)$ (this is called the mode-function describing the confinement of the pulse in the transverse direction and, in general, consists of two parts corresponding to the right and left polarizations). We further assume that the fiber maintains the polarization (a polarization preserving fiber). Then, the solution to (3.42) may be written as

$$\tilde{\boldsymbol{E}}_l^{(1)}(r,\theta,\xi,\tau) = \begin{cases} q_1^{(1)}(\xi,\tau)\,\tilde{\boldsymbol{U}}(r,\theta) & \text{for } l=1 \\ 0 & \text{for } l \neq \pm 1 \end{cases} \quad . \tag{3.43}$$

Here, the coefficient $q_1^{(1)}(\xi,\tau)$ with $q_{-1}^{(1)} = q_1^{(1)*}$ is a complex, scalar function satisfying certain equations determined by the higher order equation of (3.39). From the expression $L_1\tilde{\boldsymbol{U}} = 0$, the inner product $(\tilde{\boldsymbol{U}}, L_1\tilde{\boldsymbol{U}}) = 0$ gives the linear dispersion relation $k_0 = k_0(\omega_0)$,

$$\begin{aligned} \frac{1}{4}\,k_0^2\,S_0 =& \frac{\omega_0^2}{c^2}\,(\tilde{\boldsymbol{U}}, n_0^2\tilde{\boldsymbol{U}}) + (\tilde{\boldsymbol{U}}, L_0\tilde{\boldsymbol{U}}) \\ &+ \mathrm{i}k_0 \int (U_z\nabla_\perp\cdot\tilde{\boldsymbol{U}}^* - U_z^*\nabla_\perp\cdot\tilde{\boldsymbol{U}})\,dS \quad , \end{aligned} \tag{3.44}$$

where $L_0 = L_l\,(l=0)$, $n_0 = \sqrt{\chi_0^{(0)}}$ is the refractive index, and we have assumed the normalization for $\tilde{\boldsymbol{U}}$ to be $\int(|U_r|^2 + |U_\theta|^2)\,dS = S_0/4$.

At order ε^2, we have

$$L_l\,\tilde{\boldsymbol{E}}_l^{(2)} = \mathrm{i}\left\{-\frac{\partial L_l}{\partial\omega_l} - \left(\frac{1}{v_g} - \frac{\partial k_0}{\partial\omega_0}\right)\frac{\partial}{\partial k_l}\cdot\left(L_l - \frac{\omega_l^2}{c^2}\,\chi_l^{(0)}\right)\right\}\frac{\partial\tilde{\boldsymbol{E}}_l^{(1)}}{\partial\tau}\,, \tag{3.45}$$

from which we obtain $\tilde{\boldsymbol{E}}_l^{(2)} = 0$ if $l \neq \pm 1$. In the case where $l = 1$, it is necessary that the inhomogeneous equation (3.45) satisfies the compatibility condition (the condition required for (3.45) to be solvable, which is known as the Fredholm alternative)

$$(\tilde{\boldsymbol{U}}, L_1\,\tilde{\boldsymbol{E}}_l^{(2)}) = 0 \quad . \tag{3.46}$$

This gives the group velocity v_g in terms of the linear dispersion relation (3.44)

$$\frac{1}{v_g} = \frac{\partial k_0}{\partial\omega_0} \quad , \tag{3.47}$$

and, for $l = 1$ (3.45) becomes

$$L_1\,\tilde{\boldsymbol{E}}_l^{(2)} = -\mathrm{i}\,\frac{\partial L_1}{\partial\omega_0}\,\frac{\partial\tilde{\boldsymbol{E}}_l^{(1)}}{\partial\tau} = -\mathrm{i}\,\frac{\partial L_1}{\partial\omega_0}\,\frac{\partial q_1^{(1)}}{\partial\tau}\,\tilde{\boldsymbol{U}} \quad . \tag{3.48}$$

From (3.41) for $l = 1$, the solution of (3.48) may be found in the form

$$\tilde{\boldsymbol{E}}_l^{(2)} = \mathrm{i}\,\frac{\partial q_1^{(1)}}{\partial\tau}\,\frac{\partial\tilde{\boldsymbol{U}}}{\partial\omega_0} + q_l^{(2)}\tilde{\boldsymbol{U}} \quad , \tag{3.49}$$

where $q_1^{(2)} = q_1^{(2)}(\xi,\tau)$ with $q_{-1}^{(2)} = q_1^{(2)*}$ is a scalar function to be determined in the higher order equation. As we will see later, the first term in (3.49) represents the effect of wave guide dispersion in the coefficient of the nonlinear dispersion terms in the nonlinear Schrödinger equation.

At order ε^3, we have

$$L_l\,\tilde{\boldsymbol{E}}_l^{(3)} \begin{cases} = 0 & \text{if } l \neq \pm 1\,,\ \pm 3 \\ = -\frac{27\omega_0^2}{c^2}\chi_{111}^{(2)}\,q_1^{(1)3}\,(\tilde{\boldsymbol{U}}\cdot\tilde{\boldsymbol{U}})\,\tilde{\boldsymbol{U}} & \text{if } l = 3 \\ = \mathrm{i}\,\frac{\partial L_1}{\partial\omega_0}\,\frac{\partial\tilde{\boldsymbol{E}}_1^{(2)}}{\partial\tau} + \frac{1}{2}\,\frac{\partial^2 L_1}{\partial\omega_0^2}\,\frac{\partial^2\tilde{\boldsymbol{E}}_1^{(1)}}{\partial\tau^2} & \\ \quad + \left(\mathrm{i}\,\frac{\partial q_1^{(1)}}{\partial\xi} - \frac{1}{2}\,\frac{\partial^2 k_0}{\partial\omega_0^2}\,\frac{\partial^2 q_1^{(1)}}{\partial\tau^2}\right) & \\ \quad \times\left\{\frac{\partial}{\partial k_0}\left(L_1 - \frac{\omega_0^2 n_0^2}{c^2}\right)\right\}\tilde{\boldsymbol{U}} & \\ \quad - |q_1^{(1)}|^2\,q_1^{(1)}\,\frac{\omega_0^2}{c^2}\,\tilde{\boldsymbol{F}}\,(\tilde{\boldsymbol{U}},\tilde{\boldsymbol{U}}^*;\chi^{(2)}) & \text{if } l = 1 \end{cases} \tag{3.50}$$

where the column vector $\tilde{\boldsymbol{F}}$ is given by $\tilde{\boldsymbol{F}}\,(\tilde{\boldsymbol{U}},\tilde{\boldsymbol{U}}^*;\chi^{(2)}) = \chi_{-111}^{(2)}(\tilde{\boldsymbol{U}}^*\cdot\tilde{\boldsymbol{U}})\,\tilde{\boldsymbol{U}} + \chi_{1-11}^{(2)}(\tilde{\boldsymbol{U}}\cdot\tilde{\boldsymbol{U}}^*)\,\tilde{\boldsymbol{U}} + \chi_{11-1}^{(2)}(\tilde{\boldsymbol{U}}\cdot\tilde{\boldsymbol{U}})\,\tilde{\boldsymbol{U}}^*$. (Note that if $\tilde{\boldsymbol{U}}$ is real and $\chi^{(2)}$ is symmetric, $\tilde{\boldsymbol{F}} = \chi^{(2)}(\tilde{\boldsymbol{U}}\cdot\tilde{\boldsymbol{U}})\,\tilde{\boldsymbol{U}}$.) From (3.50), one can obtain the solutions $\tilde{\boldsymbol{E}}_l^{(3)} = 0$ for $l \neq \pm 1$ or ± 3, and since l_3 does not have an eigenmode,

$$\tilde{\boldsymbol{E}}_3^{(3)} = -\frac{27\omega_0^2}{c^2}\,q_1^{(1)3}\,L_3^{-1}\left\{\chi_{111}^{(2)}(\tilde{\boldsymbol{U}}\cdot\tilde{\boldsymbol{U}})\,\tilde{\boldsymbol{U}}\right\} \quad , \tag{3.51}$$

which is the harmonic generated by the nonlinearity. For $l = 1$, we again require the compatibility condition

$$\left(\tilde{\boldsymbol{U}}\,L_1\,\tilde{\boldsymbol{E}}_1^{(3)}\right) = 0 \quad , \tag{3.52}$$

from which we obtain the nonlinear Schrödinger equation for $q_1^{(1)}(\xi,\tau)$,

$$\mathrm{i}\,\frac{\partial q_1^{(1)}}{\partial\xi} - \frac{1}{2}\,\frac{\partial^2 k_0}{\partial\omega_0^2}\,\frac{\partial^2 q_1^{(1)}}{\partial\tau^2} + \nu\,|q_1^{(1)}|^2\,q_1^{(1)} = 0 \quad . \tag{3.53}$$

Here, the Kerr coefficient ν is a positive real number given by

$$\nu = \frac{2\omega_0^2}{\tilde{k}_0\,c^2\,S_0}\,(\tilde{\boldsymbol{U}},\tilde{\boldsymbol{F}}(\tilde{\boldsymbol{U}},\tilde{\boldsymbol{U}}^*;\chi^{(2)})) \quad , \tag{3.54}$$

where $\tilde{k}_1 = k_1 - (2\mathrm{i}/S_0)\int(U_z\nabla_\perp\cdot\tilde{\boldsymbol{U}}^* - U_z^*\nabla_\perp\cdot\tilde{\boldsymbol{U}})\,d\boldsymbol{S}$. It is worth noting that the explicit form given in (3.49) for $\tilde{\boldsymbol{E}}_1^{(2)}$ is unnecessary in the calculation of the compatibility condition (3.52), and that (3.52) can be obtained directly from the equations for $\tilde{\boldsymbol{E}}_l^{(1)}$ and $\tilde{\boldsymbol{E}}_1^{(2)}$, i.e. (3.42) and (3.48).

In order to see the effect of the higher order terms, one needs to find the equation for $q_1^{(2)}$ in (3.49). For this purpose, we have, at order ε^4, the expression $L_l\,\tilde{\boldsymbol{E}}_l^{(4)}$ for $l = 1$,

$$
\begin{aligned}
L_1 \tilde{\boldsymbol{E}}_l^{(4)} = & -\mathrm{i}\,\frac{\partial L_1}{\partial \omega_0}\,\frac{\partial \tilde{\boldsymbol{E}}_l^{(3)}}{\partial \tau} + \frac{1}{2}\,\frac{\partial^2 L_1}{\partial \omega_0^2}\,\frac{\partial^2 \tilde{\boldsymbol{E}}_1^{(2)}}{\partial \tau^2} + \frac{\mathrm{i}}{6}\,\frac{\partial^3 L_1}{\partial \omega_0^3}\,\frac{\partial^3 \tilde{\boldsymbol{E}}_l^{(1)}}{\partial \tau^3} \\
& + \left\{ \frac{\partial}{\partial k_0} \left(L_1 - \frac{\omega_0^2 n_0^2}{c^2} \right) \right\} \\
& \times \left(\mathrm{i}\,\frac{\partial \tilde{\boldsymbol{E}}_l^{(2)}}{\partial \xi} - \frac{1}{2}\,\frac{\partial^2 k_0}{\partial \omega_0^2}\,\frac{\partial^2 \tilde{\boldsymbol{E}}_l^{(2)}}{\partial \tau^2} - \frac{\mathrm{i}}{6}\,\frac{\partial^3 k_0}{\partial \omega_0^3}\,\frac{\partial^3 \tilde{\boldsymbol{E}}_1^{(1)}}{\partial \tau^3} \right) \qquad (3.55) \\
& + \mathrm{i} \left\{ \frac{\partial^2}{\partial \omega_0\, \partial k_0} \left(L_1 - \frac{\omega_0^2 n_0^2}{c^2} \right) \right\} \frac{\partial}{\partial \tau} \left(\mathrm{i}\,\frac{\partial \tilde{\boldsymbol{E}}_l^{(1)}}{\partial \xi} - \frac{1}{2}\,\frac{\partial^2 k_0}{\partial \omega_0^2}\,\frac{\partial^2 \tilde{\boldsymbol{E}}_l^{(1)}}{\partial \tau^2} \right) \\
& - \frac{1}{c^2} \sum_{l_1+l_2+l_3=1} \left[\omega_0^2 \sum_{i+j+k=4} \chi^{(2)}_{l_1 l_2 l_3} \left(\tilde{\boldsymbol{E}}_{l_1}^{(i)} \cdot \tilde{\boldsymbol{E}}_{l_2}^{(j)} \right) \tilde{\boldsymbol{E}}_{l_3}^{(k)} \right. \\
& \left. + \mathrm{i} \sum_{i=1}^{3} \left\{ \left(2\omega_0 + \omega_0^2 \frac{\partial}{\partial \omega_{l_1}} \right) \chi^{(2)}_{l_1 l_2 l_3} \right\} \frac{\partial}{\partial \tau_i} \left(\tilde{\boldsymbol{E}}_{l_1}^{(1)} \cdot \tilde{\boldsymbol{E}}_{l_2}^{(1)} \right) \tilde{\boldsymbol{E}}_{l_3}^{(1)} \right] \quad .
\end{aligned}
$$

Using (3.42), (3.48), (3.50) and the remark following (3.54), one can calculate the compatibility condition for (3.55), i.e. $(\tilde{\boldsymbol{U}}, L_1 \tilde{\boldsymbol{E}}_l^{(4)}) = 0$. The resulting equation for $q_l^{(2)}$ is

$$
\begin{aligned}
& \mathrm{i}\,\frac{\partial q_1^{(2)}}{\partial \xi} - \frac{1}{2}\,\frac{\partial^2 k_0}{\partial \omega_0^2}\,\frac{\partial^2 q_1^{(2)}}{\partial \tau^2} + 2\nu\, |q_1^{(1)}|^2\, q_1^{(2)} + \nu\, q_1^{(1)2}\, q_1^{(2)*} \\
& - \frac{\mathrm{i}}{6}\,\frac{\partial^3 k_0}{\partial \omega_0^3}\,\frac{\partial^3 q_1^{(1)}}{\partial \tau^3} + \mathrm{i} a_1 \frac{\partial}{\partial \tau} (|q_1^{(1)}|^2\, q_1^{(1)}) + \mathrm{i} a_2\, q_1^{(1)} \frac{\partial}{\partial \tau} |q_1^{(1)}|^2 = 0 \quad . \qquad (3.56)
\end{aligned}
$$

Here, the coefficient of the nonlinear dispersion terms a_1 and a_2 is

$$
a_1 = \frac{\partial \nu}{\partial \omega_0} \qquad (3.57a)
$$

$$
a_2 = \frac{2\omega_0^2}{\tilde{k}_1\, c^2\, S_0} (\tilde{\boldsymbol{U}}, \tilde{\boldsymbol{G}}(\tilde{\boldsymbol{U}}, \tilde{\boldsymbol{U}}^*; \chi^{(2)})) \quad , \qquad (3.57b)
$$

where

$$
\begin{aligned}
\tilde{\boldsymbol{G}} = & \frac{\partial \tilde{\boldsymbol{F}}}{\partial \omega_0} + \chi^{(2)}_{-111} (\dot{\tilde{\boldsymbol{U}}}^* \cdot \tilde{\boldsymbol{U}}) \tilde{\boldsymbol{U}} - \chi^{(2)}_{-\dot{1}11} (\tilde{\boldsymbol{U}}^* \cdot \tilde{\boldsymbol{U}}) \tilde{\boldsymbol{U}} \\
& + \chi^{(2)}_{1-11} (\tilde{\boldsymbol{U}} \cdot \dot{\tilde{\boldsymbol{U}}}^*) \tilde{\boldsymbol{U}} - \chi^{(2)}_{1-\dot{1}1} (\tilde{\boldsymbol{U}} \cdot \tilde{\boldsymbol{U}}^*) \tilde{\boldsymbol{U}} \\
& + \chi^{(2)}_{11-1} (\tilde{\boldsymbol{U}} \cdot \tilde{\boldsymbol{U}}) \dot{\tilde{\boldsymbol{U}}}^* - \chi^{(2)}_{11-\dot{1}} (\tilde{\boldsymbol{U}} \cdot \tilde{\boldsymbol{U}}) \tilde{\boldsymbol{U}}^*
\end{aligned}
$$

with $\dot{\tilde{\boldsymbol{U}}} = \partial \tilde{\boldsymbol{U}} / \partial \omega_0$. A "dot" over the subscript indicates the partial derivative with respect to that component of frequency, i.e., $\chi_{\dot{l}_1 l_2 l_3} = \partial \chi_{l_1 l_2 l_3} / \partial \omega_{l_1}$, and so on. The equation for $\tilde{\boldsymbol{E}}_1$ can not be separated in each order of ε, and one should use the following equation for the combined variable $q_1 = \varepsilon\, q_1^{(1)} + \varepsilon^2\, q_1^{(2)}$ (which is given by the projection of $\tilde{\boldsymbol{E}}_1$ onto $\tilde{\boldsymbol{U}}$, i.e. $q_1 = 4\int \tilde{\boldsymbol{E}}_1 \cdot \tilde{\boldsymbol{U}}\, d\boldsymbol{S}/S_0$),

$$\mathrm{i}\,\frac{\partial q_1}{\partial \xi} - \frac{1}{2}\,\frac{\partial^2 k_1}{\partial \omega_0^2}\,\frac{\partial^2 q_1}{\partial \tau^2} + \tilde{\nu}\,|q_1|^2 q_1 - \varepsilon\,\frac{\mathrm{i}}{6}\,\frac{\partial^3 k_1}{\partial \omega_0^3}\,\frac{\partial^3 q_1}{\partial \tau^3}$$
$$+\,\varepsilon\,\mathrm{i}\tilde{a}_1\,\frac{\partial}{\partial \tau}\left(|q_1|^2 q_1\right) + \varepsilon\,\mathrm{i}\tilde{a}_2 q_1\,\frac{\partial}{\partial \tau}\left(|q_1|^2\right) = O(\varepsilon^3) \quad , \tag{3.58}$$

with $\tilde{\nu} = \nu/\varepsilon^2$, $\tilde{a}_1 = a_1/\varepsilon^2$, $\tilde{a}_2 = a_2/\varepsilon^2$. Equation (3.58) is the equation we seek for the envelope function q_1.

The coefficients of the nonlinear envelope equation (3.58) may be simplified for a weakly guided fiber. From here on, we remove from ω and k the subscripts 0 which were used to designate the carrier frequency and wave number. In order to derive the simplified expression, we first calculate the approximated mode function $\tilde{\boldsymbol{U}}$ satisfying (3.42), $L_1\tilde{\boldsymbol{U}} = 0$, assuming weak guiding, i.e.

$$|\lambda\, \nabla_\perp\, \ln n_0| \simeq O(\delta) \ll 1 \quad , \tag{3.59}$$

where λ is the wavelength of the carrier, and then evaluate the coefficients given by (3.44), (3.54) and (3.57).

We write the electric field, $\hat{\boldsymbol{E}} = 2q_1\,\boldsymbol{U}$, (with the unit vector $\hat{z}$) as

$$\hat{\boldsymbol{E}} = 2q_1\,\boldsymbol{U} = \nabla\phi \times \hat{z} + \boldsymbol{V} \qquad \text{with} \quad (\nabla\phi \times \hat{z}) \cdot \boldsymbol{V} = 0 \quad . \tag{3.60}$$

(Note that if $\boldsymbol{V} = 0$, an exact TE mode is implied.) Then, (3.42) becomes

$$\hat{z} \times \nabla_\perp \left\{ (\nabla_\perp^2 - k^2 + \frac{\omega^2}{c^2}\, n_0^2)\phi \right\} - \frac{\omega^2}{c^2}\, \phi \hat{z} \times \nabla_\perp n_0^2 - L_1\, \tilde{\boldsymbol{V}} = 0 \quad , \tag{3.61}$$

where the column vector $L_1\, \tilde{\boldsymbol{V}}$ is considered to be the usual vector. Using the small parameter δ in (3.59), we solve (3.61) by the perturbation expansion, i.e., $k = k^{(0)} + \delta k^{(1)} + \ \ldots$ and $V = \delta V^{(1)} + \ \ldots$. Then, in leading order, we have

$$\nabla_\perp^2\, \phi - k^{(0)2}\, \phi + \frac{\omega^2}{c^2}\, n_0^2\, \phi = 0 \quad , \tag{3.62}$$

where we have assumed that $|\phi| \to 0$ as $\| \boldsymbol{x}_\perp \| \to \infty$, i.e. the bound state. Here, the leading order eigenvalue $k^{(0)}$ is obtained by

$$k^{(0)2} = \frac{(\omega/c)^2 \int |\phi|^2\, n_0^2\, dS - \int |\nabla_\perp \phi|^2\, dS}{\int |\phi|^2\, dS} \quad . \tag{3.63}$$

At the order of δ, we have

$$L_1\, \tilde{\boldsymbol{V}} = -2k^{(0)} k^{(1)} \hat{z} \times \nabla_\perp \phi - \frac{\omega^2}{c^2}\, \phi\, (\hat{z} \times \nabla_\perp n_0^2)\, \delta^{-1} \quad . \tag{3.64}$$

From the compatibility condition (see the statement preceding (3.46)), i.e., $(\tilde{\boldsymbol{U}}, L_1\, \tilde{\boldsymbol{V}}) = 0$, we obtain the equation for $k^{(1)}$,

$$2\delta k^{(0)}k^{(1)} = -\frac{(\omega/c)^2 \int \phi^*(\nabla_\perp \phi \cdot \nabla_\perp n_0^2)\, dS}{\int |\nabla_\perp \phi|^2\, dS} + O(\delta^2) \quad , \tag{3.65}$$

where we have used the condition that (3.62) has only one bound state, i.e. the monomode fiber. From (3.62), (3.63) and (3.65), we obtain the approximated eigenvalue k, i.e., the linear dispersion relation,

$$\begin{aligned} k^2 &= k^{(0)2} + 2\delta\, k^{(0)}\, k^{(1)} + O(\delta^2) \\ &= \frac{(\omega/c)^2 \int |\nabla_\perp \phi|^2 n_0^2\, dS - \int |\nabla_\perp^2 \phi|^2\, dS}{\int |\nabla_\perp \phi|^2\, dS} + O(\delta^2) \quad . \end{aligned} \tag{3.66}$$

We note that (3.66) can be obtained from $(\tilde{U}, L_1 \tilde{U}) = 0$ with $2q_1 \tilde{U} = \nabla_\perp \phi \times \hat{z}$, where ϕ satisfies (3.62). Thus, we can assume the TE mode up to order δ, and the effect of an axial component of the electric field appears at order δ^2.

Let us now evaluate the first term in the coefficient of the nonlinear dispersion, ν, (3.54). When (3.60) is used in (3.54), the coefficient ν becomes

$$\begin{aligned} \nu &= \frac{\omega^2 S_0}{8kc^2} \frac{3 \int \chi_2 |\nabla_\perp \phi|^4\, dS}{\left(\int |\nabla_\perp \phi|^2\, dS\right)^2} + O(\delta^2) \\ &= \frac{\omega^2}{kc^2\, S_0\, E_0^4} \int n_0 n_2 |\nabla_\perp \phi|^4\, dS + O(\delta^2) \quad , \end{aligned} \tag{3.67}$$

where we have assumed the medium to be isotropic and symmetric in the instantaneous nonlinear response, i.e., $\chi^{(2)}_{-111} = \chi^{(2)}_{1-11} = \chi^{(2)}_{11-1} = \chi_2$. This nonlinear dielectric constant χ_2 may be given by the following form of the real nonlinear response function $\chi^{(2)}$ in (3.33):

$$\chi_2(\omega) = \int_{-\infty}^{0} dt_1 \int_{-\infty}^{0} dt_2 \int_{-\infty}^{0} dt_3 \chi^{(2)}(t_1, t_2, t_3)\, \mathrm{e}^{-\mathrm{i}\omega(t_1 - t_2 - t_3)} \quad , \tag{3.68}$$

where the imaginary part of χ_2 may be negligible for most practical fibers. Equation (3.67) can be evaluated given the eigenfunction ϕ. In particular, for a dielectric wave guide with step dielectric profile, ϕ is given by a Bessel function and ν is reduced to the value obtained earlier [3.8].

Given these assumptions, the coefficients of the higher order nonlinear dispersion and dissipation terms a_1 and a_2 are obtained in the same way. Using (3.57a) and (3.67),

$$\begin{aligned} a_1 &= \frac{\partial \nu}{\partial \omega} \\ &= \frac{\partial}{\partial \omega}\left[\frac{\omega^2}{kc^2\, S_0\, E_0^4} \int n_0 n_2 |\nabla_\perp \phi|^4\, dS\right] + O(\delta^2) \quad . \end{aligned} \tag{3.69}$$

Similarly, the coefficient a_2 is obtained from (3.57b),

$$a_2 = \frac{\omega^2}{kc^2\, S_0\, E_0^4} \int \left[n_0 n_2 \frac{\partial |\nabla\phi|^4}{\partial\omega} + \tfrac{3}{4}\left(\chi_1^{(2)} - \chi_{-1}^{(2)}\right) |\nabla\phi|^4 \right] dS + O(\delta^2) \quad , \tag{3.70}$$

where

$$\chi_1^{(2)} \equiv \chi_{-1\mathrm{i}1}^{(2)} = \chi_{-11\mathrm{i}}^{(2)} = \chi_{\mathrm{i}-11}^{(2)} = \chi_{1-1\mathrm{i}}^{(2)} = \chi_{\mathrm{i}1-1}^{(2)} = \chi_{1\mathrm{i}-1}^{(2)}$$

and

$$\chi_{-1}^{(2)} \equiv \chi_{-\mathrm{i}11}^{(2)} = \chi_{1-\mathrm{i}1}^{(2)} = \chi_{11-\mathrm{i}}^{(2)}$$

are derivatives of $\chi^{(2)}$ with respect to ω and $-\omega$. From the definitions of $\chi_{-1}^{(2)}$ and $\chi_1^{(2)}$, we have the formula for $\chi_1^{(2)} - \chi_{-1}^{(2)}$ in terms of the real nonlinear response function $\chi^{(2)}$ in (3.2),

$$\chi_1^{(2)}(\omega) - \chi_{-1}^{(2)}(\omega) = \frac{\mathrm{i}}{2} \int_{-\infty}^{0} dt_1 \int_{-\infty}^{0} dt_2 \int_{-\infty}^{0} dt_3 \times \chi^{(2)}(t_1, t_2, t_3)\,(2t_1 - t_2 - t_3)\,\mathrm{e}^{-\mathrm{i}\omega(t_1 - t_2 - t_3)} \quad , \tag{3.71}$$

which is imaginary. We note that the term a_2 can be obtained by the retarded nonlinear response which has the form

$$q_1(t) \int_{-\infty}^{t} ds\, f(t-s)\, |q_1(s)|^2 \quad ,$$

assuming short delays (this assumption is consistent with the quasi-monochromatic approximation). As has been shown by *Gordon* [3.9], this term corresponds to the retarded Raman effect in which the higher frequency components of the soliton spectrum pump energy to the lower frequency components. Consequently, this term produces a nonlinear dissipation and, as will be discussed in Sect. 8.1, leads to a down-shift of the carrier frequency of the soliton.

4. Envelope Solitons

In this chapter we describe various properties of envelope solitons.

4.1 Soliton Solutions and the Results of Inverse Scattering

Following the method of *Gardner* et al.[4.2] and of *Lax* [4.3], Zakharov and Shabat have discovered that the eigenvalue λ of the Dirac-type eigenvalue equations,

$$L\psi = \lambda\psi \quad , \tag{4.1}$$

$$\psi = \begin{pmatrix} \psi_1 \\ \psi_2 \end{pmatrix} \tag{4.2}$$

with

$$L = \mathrm{i}\begin{pmatrix} 1+\beta & 0 \\ 0 & 1-\beta \end{pmatrix}\frac{\partial}{\partial x} + \begin{bmatrix} 0 & u^* \\ u & 0 \end{bmatrix}$$

becomes time-invariant if u evolves in accordance with the nonlinear Schrödinger equation of the form,

$$\mathrm{i}\frac{\partial u}{\partial t} + \frac{\partial^2 u}{\partial x^2} + Q\,|u|^2\,u = 0 \quad , \tag{4.3}$$

where $\beta^2 = 1 - 2/Q = \text{const}$, and the time evolution of the eigenfunction ψ is given by

$$\mathrm{i}\frac{\partial \psi}{\partial t} = A\,\psi \quad , \tag{4.4}$$

where

$$A = -\beta\begin{bmatrix} 1 & 0 \\ 0 & 1 \end{bmatrix}\frac{\partial^2}{\partial x^2} + \begin{bmatrix} \frac{|u|^2}{1+\beta} & \mathrm{i}\frac{\partial u^*}{\partial x} \\ -\mathrm{i}\frac{\partial u}{\partial x} & \frac{-|u|^2}{1-\beta} \end{bmatrix} \quad . \tag{4.5}$$

Once the structure of the eigenvalue equation which satisfies the Lax criteria has been discovered, one can apply the inverse scattering technique to obtain the time evolution of the potential q, and the nonlinear Schrödinger equation can then be solved for a localized initial condition. As in the case of the KdV equation, the time invariance of the eigenvalues provides those properties of solitons which are created from the initial condition in terms of the eigenvalues of the initial potential shape. For the nonlinear Schrödinger equation (3.15), the appropriate structure of the eigenvalue equation becomes [4.4]

$$\mathrm{i}\,\frac{\partial \psi_1}{\partial T} + q_0(T)\,\psi_2 = \zeta\,\psi_1 \tag{4.6}$$

$$-\mathrm{i}\,\frac{\partial \psi_2}{\partial T} - q_0^*(T)\,\psi_1 = \zeta\,\psi_2 \quad . \tag{4.7}$$

If we write the eigenvalue of this equation as

$$\zeta_n = \frac{\kappa_n + \mathrm{i}\eta_n}{2} \quad , \tag{4.8}$$

the n soliton solutions which arise from this initial wave form are a generalization of (3.31), and are given by

$$q(T,Z) = \sum_{j=1}^{N} \eta_j \operatorname{sech} \eta_j(T + \kappa_j Z - \theta_{0j}) \times \exp\left\{-\mathrm{i}\,\kappa_j T + \frac{\mathrm{i}}{2}(\eta_j^2 - \kappa_j^2)Z - \mathrm{i}\,\sigma_{0j}\right\} \quad . \tag{4.9}$$

We note that the amplitude and speed of the soliton are characterized by the imaginary and real parts of the eigenvalue (4.8).

For example, if we approximate the input pulse shape of a mode-locked laser as

$$q_0(T) = A \operatorname{sech} T \quad , \tag{4.10}$$

the eigenvalues of (4.7) are obtained analytically and the number of eigenvalues N is given by [4.4]

$$A - \tfrac{1}{2} < N \le A + \tfrac{1}{2} \quad , \tag{4.11}$$

where the corresponding eigenvalues are imaginary and given by

$$\zeta_n = \mathrm{i}\,\frac{\eta_n}{2} = \mathrm{i}\left(A - n + \tfrac{1}{2}\right) \quad , \qquad n = 1, 2, \ldots, N \quad . \tag{4.12}$$

If A is exactly equal to N, the solution can be obtained in terms of N solitons and their amplitudes are then given by

$$\eta_n = 2(N-n) + 1 = 1, 3, 5, \ldots, (2N-1) \quad . \tag{4.13}$$

We should note here that in this particular case, all the eigenvalues are purely imaginary. Consequently, $\kappa_n = 0$, and all the soliton velocities in the frame of reference of the group velocity are 0.

While the speed of the solitons in the KdV equation are proportional to the amplitude, those in the Schrödinger equation have no such dependence. In general, if the input pulse shape is symmetric, as in this example of sech T, the

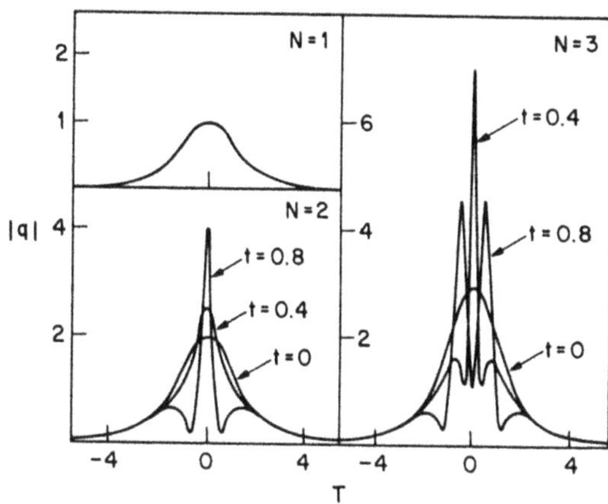

Fig. 4.1. Higher order solitons which emerge from initial condition of $N\,\mathrm{sech}\,T$ [4.4]

eigenvalues of (4.6) and (4.7) can be shown to be purely imaginary, and the output solitons propagate at exactly the same speed [4.5]. When a number of solitons propagate at the same speed, the superimposed pulse shape oscillates due to the phase interference among the solitons, as shown in Fig. 4.1.

If we remember that the eigenvalues are given by (4.12), we can obtain the amplitude of the output soliton when the initial amplitude is slightly different from that which corresponds to the one-soliton solution. For example, if we write the input pulse shape as

$$q_0(T) = (1+\Delta)\,\mathrm{sech}\,T \quad , \tag{4.14}$$

(4.12) gives $\eta = 1+2\Delta$ for the value of $A = 1+\Delta$. From this, the output soliton shape is given by

$$q(T) = (1+2\Delta)\,\mathrm{sech}\,(1+2\Delta)T \quad . \tag{4.15}$$

Since the soliton energy is given by

$$\varepsilon = \int_{-\infty}^{\infty} |q|^2\,dT \quad , \tag{4.16}$$

the difference in energy between the input and output becomes

$$\Delta\varepsilon = \int_{-\infty}^{\infty} (1+\Delta)^2\,\mathrm{sech}^2\,T\,dT - \int_{-\infty}^{\infty} (1+2\Delta)^2\,\mathrm{sech}^2\,(1+2\Delta)\,T\,dT = \Delta^2 \quad , \tag{4.17}$$

i.e., the soliton energy is smaller by a factor Δ^2 as compared to the input pulse energy. This means that if the input amplitude is not exactly an integer N, part of the energy in the input pulse is transferred to other waves, namely, to a linear dispersive wave which does not form a soliton.

4.2 Soliton Periods

N solitons which propagate at the same speed propagate together with phase interactions, as shown in Fig. 4.1 [4.4].

In general, the periodicity of this oscillation is given by the lowest common beat frequency,

$$Z_0 = \frac{2\pi}{|\omega_i - \omega_j|} \quad , \tag{4.18}$$

where $\omega_j = \eta_j^2/2$, η_j being the imaginary part of the eigenvalues of the initial pulse shape [4.5]. In particular, when the pulse shape is given by (4.10), the period of oscillation of n solitons reduces to the simple form,

$$Z_0 = \frac{\pi}{2} \quad . \tag{4.19}$$

The quantity Z_0 is often referred to as a soliton period.

4.3 Conservation Quantities of the Nonlinear Schrödinger Equation

The nonlinear Schrödinger equation is integrable owing to the fact that it has an infinite number of conserved quantities. We write here the three lowest conserved quantities;

$$C_1 = \int_{-\infty}^{\infty} |q(Z,T)|^2 \, dT \tag{4.20}$$

$$C_2 = \int_{-\infty}^{\infty} \left(q^* \frac{\partial q}{\partial T} - q \frac{\partial q^*}{\partial T} \right) dT \tag{4.21}$$

and

$$C_3 = \int_{-\infty}^{\infty} \left(\left| \frac{\partial q}{\partial T} \right|^2 - |q|^4 \right) dT \quad . \tag{4.22}$$

We note that the one-soliton solution (3.31) is obtained by minimizing C_3, the constraints C_1 and C_2 being constant.

5. Solitons in Optical Fibers

In the preceding chapter it was shown that, under ideal conditions, a light wave in a dielectric material would form an envelope soliton. However, in a real three-dimensional fiber, the light wave propagates with finite loss. In this chapter, we discuss the practical conditions needed for an optical pulse to become a soliton in a dielectric waveguide.

5.1 Effect of Fiber Loss

Let us first consider the effect of fiber loss. If the fiber loss is too large, the amplitude of the light wave decreases as it propagates and the pulse may be dissipated before it forms a soliton. Hence, let us study the conditions under which a soliton is formed when fiber loss is present [5.1–3].

If we denote the loss rate per unit length by γ, the effect of the fiber loss can be incorporated into (3.13) by adding a term $-\mathrm{i}\gamma E$ to the right-hand side,

$$\mathrm{i}\frac{\partial E}{\partial \xi}-\frac{k''}{2}\frac{\partial^2 E}{\partial \tau^2}+g\frac{|E|^2 E}{\varepsilon^2}=-\frac{\mathrm{i}\gamma E}{\varepsilon^2} \quad , \tag{5.1}$$

or, in the normalized form,

$$\mathrm{i}\frac{\partial q}{dZ}+\frac{1}{2}\frac{\partial^2 q}{\partial T^2}+|q|^2 q=-\mathrm{i}\Gamma q \quad , \tag{5.2}$$

where

$$\Gamma=\frac{\gamma\lambda}{\varepsilon^2} \quad . \tag{5.3}$$

From this, we see that a soliton is formed on condition that the nonlinear term on the left-hand side, $\pi n_2|E|^2/\lambda$, is larger than γ. For example, if we take as a nominal value of the electric field, 10^6 V/m, the nonlinear term at wavelengths of $1.5\,\mu$m becomes of the order of $2\times10^{-4}\,\mathrm{m}^{-1}$. From this, we can see that the loss rate γ should be smaller than $2\times10^{-4}\,\mathrm{m}^{-1}$ in order for a light wave having this electric field to become a soliton. If we write this loss rate in terms of the power loss rate in dB/km, the critical loss rate required becomes 1.7 dB/km. Since the fiber which is commercially available at present has a loss rate of less than 0.2 dB/km at wavelengths of $1.5\,\mu$m, this condition is easily satisfied. However, it took several years until the fiber loss rate became less than 1 dB/km. This is one of the reasons why the first experiment to verify the existence of an optical soliton did not take place until some seven years after the theory of optical solitons was published.

If the fiber loss is sufficiently small, one can obtain the one-soliton solution using a perturbation technique [4.1]. The result reads,

$$q(T,Z) = \eta(Z)\,\mathrm{sech}\,[\eta(Z)T]\,\exp[\mathrm{i}\,\sigma(Z)] + O(\Gamma) \quad , \tag{5.4}$$

where

$$\eta(Z) = q_0 \exp(-2\Gamma Z)$$

and

$$\sigma(Z) = \frac{q_0^2}{8\Gamma}\,[1 - \exp(-4\Gamma Z)] \quad . \tag{5.5}$$

Equation (5.4) indicates that the soliton amplitude decreases exponentially as $\exp(-2\Gamma Z)$, and the width increases exponentially as $\exp(2\Gamma Z)$. Consequently, the soliton propagates by retaining the property that the amplitude times the pulse width remains constant. Naturally, the energy of a soliton, which is given by $\int_{-\infty}^{\infty} \eta^2\,\mathrm{sech}^2\,\eta T\,dT = \eta$, also decreases in proportion to $\mathrm{e}^{-2\Gamma Z}$. Again, as will be discussed later (Sect. 6.1), if the fiber loss is compensated for by the Raman gain, there is no change in the soliton energy. The decrease in the amplitude at a rate twice as fast as a linear pulse is a consequence of the nonlinear property of a soliton. This result indicates that amplification is necessary in order for solitons to be transmitted without distortion over an extended distance.

5.2 Effect of Waveguide Property of a Fiber

The electromagnetic wave in a fiber has a vector field with three components in both the electric and magnetic fields. However, as was shown in Sect. 3.4, the one-dimensional nonlinear Schrödinger equation is still valid, provided that its coefficients are evaluated by appropriately taking into account the waveguide effects of the fiber. In particular, when the fiber has a cross sectional dimension somewhat larger then the wavelength of the light (weakly guided fiber), the coefficients are relatively simplified. Taking into account the higher order effects and the loss, the reduced equation for the normalized amplitude reads,

$$\begin{aligned} &\mathrm{i}\left(\frac{\partial q}{\partial Z} + \Gamma q\right) + \frac{1}{2}\frac{\partial^2 q}{\partial T^2} + |q|^2 q \\ &\qquad + \varepsilon\mathrm{i}\left\{\beta_1 \frac{\partial^3 q}{\partial T^3} + \beta_2 \frac{\partial}{\partial T}(|q|^2 q) + \beta_3 q \frac{\partial}{\partial T}|q|^2\right\} = 0 \quad . \end{aligned} \tag{5.6}$$

Here,

$$q = \frac{\sqrt{g\lambda}}{\varepsilon} E \tag{5.7}$$

$$T = \frac{\tau}{T_0} = \frac{\varepsilon(t - x/v_g)}{(-\lambda k'')^{1/2}} \quad , \tag{5.8}$$

$$Z = \frac{\xi}{\lambda} \quad , \tag{5.9}$$

$$T_0 = (-\lambda k'')^{1/2} \tag{5.10}$$

$$\Gamma = \frac{\gamma\lambda}{\varepsilon^2} \tag{5.3}$$

with $k''\,(= \partial^2 k/\partial\omega^2)$ given by the second derivative of the wave number k in the fiber (see Sect. 3.4),

$$k^2 = \frac{(\omega/c)^2 \int |\nabla_\perp\phi|^2 n_0^2\, dS - \int |\nabla_\perp^2 \phi|^2\, dS}{\int |\nabla_\perp\phi|^2\, dS} \quad . \tag{5.11}$$

Here, ϕ is the potential for the transverse component of electric field,

$$\boldsymbol{E}_\perp = \nabla_\perp\phi \times \hat{z} \quad ;$$

it satisfies the eigenvalue equation

$$\nabla_\perp^2\phi - k^{(0)2}\phi + \frac{\omega^2}{c^2}\, n_0^2\, \phi = 0 \tag{5.12}$$

and n_0 is the linear refractive index which is a function of the transverse coordinates $x_\perp$ and frequency ω. The integration $\int dS$ is evaluated across the section of the fiber and ϕ is normalized such that

$$\int |\nabla_\perp\phi|^2\, dS = S_0\, E_0^2 \quad , \tag{5.13}$$

where E_0 is the peak intensity of the light electric field, and S_0 is the (effective) cross section of the guide. We note that the ω derivative of (5.11) contains terms which originate from $\partial n_0/\partial\omega$ (the material dispersion) as well as from $\partial\phi/\partial\omega$ (the waveguide dispersion), and

$$g = \frac{\omega^2}{k\, c^2\, S_0\, E_0^4} \int n_0 n_2\, |\nabla\phi|^4\, dS \simeq \frac{\pi n_2}{\lambda} \quad . \tag{5.14}$$

Thus q is dimensionless. The coefficients of the higher order terms are given by,

$$\beta_1 = \frac{1}{6}\,\frac{k'''\lambda}{T_0^3} \quad , \qquad \lambda = \frac{2\pi}{k} \tag{5.15}$$

$$\beta_2 = \frac{1}{gT_0}\,\frac{\partial}{\partial\omega}\left[\frac{\omega^2}{k\, c^2\, S_0\, E_0^4} \int n_0 n_2\, |\nabla\phi|^4\, dS\right] \tag{5.16}$$

$$\beta_3 = \frac{1}{gT_0}\frac{\omega^2}{k\,c^2\,S_0\,E_0^4}\int\left[n_0n_2\frac{\partial\,|\nabla\phi|^4}{\partial\omega_0}+\frac{3}{4}\left(\chi_1^{(2)}-\chi_{-1}^{(2)}\right)|\nabla\phi|^4\right]dS\ . \quad (5.17)$$

Here β_1 represents the higher order linear dispersion, β_2 the nonlinear dispersion of the Kerr coefficient, and β_3 (which is imaginary) the nonlinear dissipation due to the Raman process in the fiber.

Other waveguide effects [5.4–6] and the effect of noise [5.7] on soliton propagation have also been studied. In addition, solitons in longer wavelength regions, where the scattering loss is theoretically expected to be much smaller, have been attracting interest [5.8]. The effect of birefringence, which is not explicitly treated in this section, influences the soliton nature for different polarizations and should be important for certain polarization preserving fibers [5.9–11].

5.3 Condition of Generation of a Soliton in Optical Fibers

Let us consider the method of generating a soliton in an optical fiber based on the properties of solitons described in this chapter.

First, we note that in the absence of loss and the higher order terms, the one-soliton solution is given by

$$q(T,Z) = \eta\,\mathrm{sech}\,\eta(T+\kappa Z-\theta_0)\ \exp\{-\mathrm{i}\kappa T+\frac{\mathrm{i}}{2}(\eta^2-\kappa^2)\,Z-\mathrm{i}\sigma_0\}\ . \quad (3.31)$$

If an optical fiber is being considered, its characteristic properties such as power loss rate δ (dB/km), group dispersion parameter D [ps/(nm km)] with $k'' = D\lambda^2/(2\pi c)$ and effective cross-section S (μm^2) are also known. Let us consider how, given these parameters, we generate a soliton in practice.

Let us first derive the relation between the peak intensity and the pulse width in terms of practical parameters. For an optical pulse in the input side of the fiber, one can use the output of a mode-locked laser. In this case, the pulse shape is close to sech t. The pulse-width τ_0 is fixed by the mode-locking, and the amplitude by the intensity of the laser oscillation.

If we write the peak power of the pulse output of the mode-locked laser as P_0, the peak electric field in the fiber E_0, and the effective cross-sectional area S, are related through,

$$P_0 = \frac{\varepsilon_0}{2}\,v_g\,E_0^2\,S\,n_0^2\ . \quad (5.18)$$

Here, $v_g\,(\simeq c/n_0)$ is the group velocity of the light wave in the fiber, $n_0\,(\simeq 1.5)$ is the effective refractive index which takes into account the waveguide effect, and $\varepsilon_0 = 8.854\times10^{-12}$ F/m is the dielectric constant of the vacuum. If we substitute these values, (5.18) becomes

$$E_0^2\,S = 5\times10^{14}\,P_0\ . \quad (5.19)$$

If we define the pulse width of the soliton as that which corresponds to half the peak power, the pulse width of an optical pulse which has an amplitude of sech$t$ shape is given by 1.76. Consequently, the relation between the soliton pulse width τ_0 and the peak electric field E_0, from (3.31), (3.16) and (3.17), is given by

$$(\pi n_2)^{1/2} E_0 \tau_0 = 1.76 (-\lambda k'')^{1/2} \quad . \tag{5.20}$$

On the other hand, if we write k'' in terms of D, we have

$$(-\lambda k'')^{1/2} = 2.3 \times 10^{-5} [\lambda(\mu\mathrm{m})]^{3/2} \left[|D| \left(\frac{\mathrm{ps}}{\mathrm{nm\,km}} \right) \right]^{1/2} (\mathrm{ps}) \quad . \tag{5.21}$$

If we use (5.18), (5.20) and (5.21), we have the relation between the pulse width τ_0 (ps) of a soliton and the required peak power P_0 (W),

$$\tau_0 \sqrt{P_0} = 9.3 \times 10^{-2} \lambda^{3/2} \sqrt{|D|\, S} \quad . \tag{5.22}$$

Here, λ is measured in μm, D in ps/(nm km), and S in $\mu\mathrm{m}^2$.

For example, if we consider a fiber with magnitude $|D| = 10\,\mathrm{ps/(nm\,km)}$ and cross-sectional area $S = 60\,\mu\mathrm{m}^2$ at wavelength $\lambda = 1.5\,\mu$m, (5.22) gives

$$\tau_0 \sqrt{P_0} = 4.2\,(\mathrm{ps\,W}^{1/2}) \quad . \tag{5.23}$$

In this case, the necessary peak input power for a soliton with $\tau_0 = 10\,\mathrm{ps}$ becomes 180 mW. However, if one uses a fiber with smaller dispersion, say, $|D| = 1$ ps/(nm km), and smaller cross-section, $S = 20\,\mu\mathrm{m}^2$, $\tau_0 \sqrt{P_0}$ becomes 0.76 and the peak power needed for a soliton with a pulse width of 10 ps is reduced to 5.8 mW. When the input power mismatches the soliton power P_0 given in this expression for the given pulse width τ_0, a soliton will still be formed, but with an amplitude and width which are different from those given in this relation [5.12]. Under these circumstances, the amplitude of the newly formed soliton is given by (4.11) in Sect. 4.1.

We now consider the minimum input power required for the formation of a soliton which is limited by fiber loss. If we write γ in (5.1) in terms of the practical loss rate of a fiber δ (dB/km),

$$\gamma\,(\mathrm{m}^{-1}) = 1.2 \times 10^{-4} \delta\,(\mathrm{dB/km}) \quad . \tag{5.24}$$

As was discussed in Sect. 5.1, if the fiber loss rate is large, a soliton can not be formed unless the sufficient soliton intensity is maintained by amplification. The condition for the formation of a soliton is given by the fact that the nonlinear term $g\,|E|^2$ in (5.1) is larger than the loss rate γ. If we write this condition using (5.23) and (5.19), we have

$$\frac{P_0\,(\mathrm{W})}{S\,(\mu\mathrm{m}^2)} > 1.9 \times 10^{-3} \delta\,(\mathrm{dB/km}) \quad . \tag{5.25}$$

For example, if the power loss rate is 0.2 dB/km and the cross-sectional area of the fiber is $60\,\mu\mathrm{m}^2$, the minimum power required for the formation of a soliton becomes 23 mW.

As will be shown, however, if one utilizes the Raman amplification, the fiber loss rate can be made effectively zero, and the minimum power requirement expressed in (5.25) then becomes unnecessary. When one chooses a particular wavelength such that the group velocity dispersion, k'', vanishes and the leading linear dispersion term becomes $\partial^3 q/\partial T^3$. Soliton-like solutions are still possible when this higher order dispersion is balanced with the Kerr nonlinear term. Such solutions have been studied numerically by *Way* et al. [5.13] and by *Agrawal* and *Potasek* [5.14].

5.4 Experiments on Generation of Optical Solitons

In order for the propagation of a soliton in an optical fiber to be verified experimentally, it is necessary to generate a short optical pulse with sufficiently large power and use a fiber which has a loss rate less than 1 dB/km. It was only in the late of 1970s that the optical fiber loss fell below 1 dB/km. For the generation of optical solitons, it is further required that the spectral width of the laser be narrower than the inverse of the pulse width in time. This requires the generation of a pulse with a narrow spectrum, known as a Fourier transform limited pulse.

For these reasons, it was seven years after the theoretical prediction that the transmission of a soliton was successfully demonstrated experimentally. In 1980, *Mollenauer* et al. [5.15] at AT&T Bell Laboratories succeeded for the first time in experimentally verifying soliton transmission in an optical fiber. They achieved this by utilizing an F^{2+} color center laser which is pumped by a Nd : YAG laser. Using a fiber with a relatively large cross-section (10^{-6} cm^2) and a length of 700 meters, they transmitted an optical pulse with a 7 ps pulse width and measured the output pulse shape by means of autocorrelation. For this particular fiber, the theoretically derived peak power for the formation of a soliton was 1.2 W. Such a large power level was chosen in order that the autocorrelation measurement could easily be made at the output side. Figure 5.1 shows this famous experimental result. Here, for different power levels at the input side, the pulse shape is measured by the autocorrelation at the output side of the fiber. It is clear from this figure that while the output pulse width increases for a power below the threshold of 1.2 W, it continuously decreases for an input power above

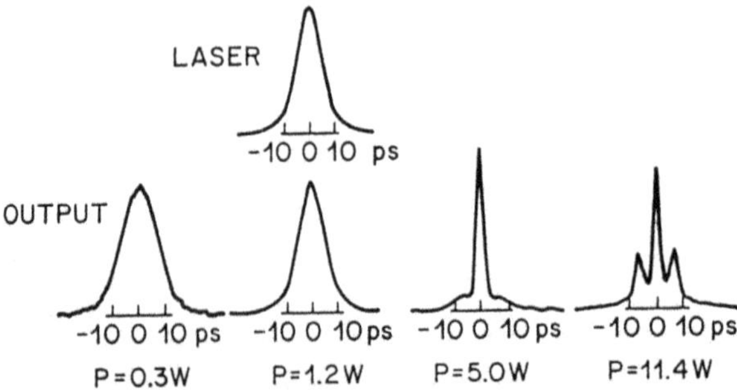

Fig. 5.1. Experimental observation of soliton formation by *Mollenauer* et al. [5.15]

1.2 W. The appearance of two peaks in the case of an input power of 11.4 W is a consequence of the phase interference of three solitons, which are generated simultaneously in this instance. This result is, in fact, consistent with the numerical calculation shown in Fig. 4.1. The periodic behaviour of the higher order solitons was confirmed by *Stolen* et al. [5.16] in a later experiment.

Due to the need to generate the Fourier transform limited pulse, most soliton experiments have been performed using the color center laser. However, solitons have been successfully generated using other types of lasers, such as a dye laser [5.17,18] or a laser diode [5.19]. In order to generate a well-defined soliton, it is necessary to control frequency chirping and to have a narrow spectrum [5.20].

6. Amplification of a Soliton — Application to the Optical Soliton Transmission System

In this chapter we discuss the amplification and reshaping of an optical soliton using, in particular, the stimulated Raman process in the same fiber. The amplification can cancel the intrinsic fiber loss, whereby a soliton can be made to propagate over an extended distance. We present the result of an experiment to verify the Raman process, a numerical simulation for a Raman amplified transmission system, an experimental result of one-soliton transmission, periodically amplified by the Raman process, over a distance of more than 6000 kms and of amplification and reshaping of solitons is an erbium doped fiber pumped by a laser diode.

6.1 Raman Amplification of Optical Solitons

Unlike a conventional linear pulse, an optical soliton is not deformed due to fiber dispersion. However, if the peak intensity of the soliton decreases as it propagates in the fiber with a loss, the pulse spreads. If the loss rate is small but finite, the soliton propagates, retaining the property that the product of the amplitude and pulse width remains constant, as shown in (5.4). Because of the increase in the pulse width, reshaping is required for the long-distance transmission of a soliton for communication purposes. Various proposals for reshaping a soliton have been put forward [6.1–6]. Among these, the most promising methods are the induced Raman amplification of the fiber itself [6.7–11] and amplification by means of an erbium doped fiber pumped by a laser diode [6.12–16].

As opposed to the case where dissipation occurs, when a soliton is amplified, its pulse width decreases in proportion to the increase in the amplitude. If the intensity of the pump wave for the Raman amplification is constant along the

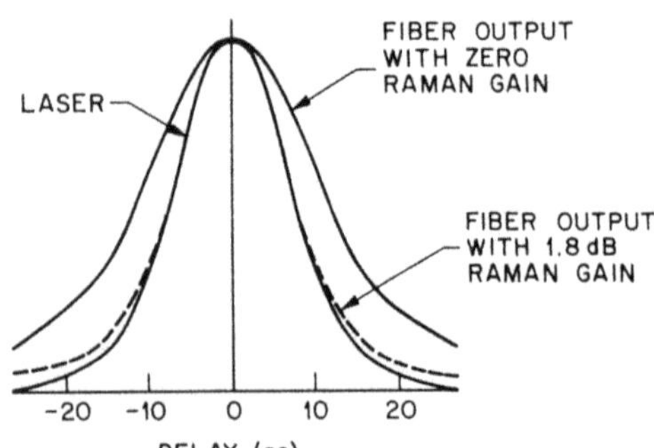

Fig. 6.1. Experimental observation of reshaping of an optical soliton by means of Raman amplification [6.9]

fiber, the gain becomes constant. Therefore, by choosing the gain so as to exactly balance the loss, the total loss rate becomes zero, which makes $\Gamma = 0$ in (5.1). When this is achieved, the soliton can propagate without any distortion, as shown in (3.31).

Figure 6.1 shows the experimental result obtained by *Mollenauer* and *Stolen* [6.9], of the reshaping of a soliton using Raman amplification in a fiber. The fiber used for the experiment had a length of 10 kms, a total loss of 1.8 dB and group dispersion of −15 ps/(nm km). A soliton with a pulse width of 10 ps and a peak power of 375 mW at a wavelength of 1.56 μm was produced using a color center laser. The Raman pump wave with a wavelength of 1.46 μm was injected from the output side of the fiber.

As shown in this figure, the soliton whose width increases in the absence of the pump wave is seen to regain its original shape when the pump intensity is adjusted to 125 mW, so that the Raman gain exactly cancels the fiber loss of 1.8 dB.

6.2 All-Optical Transmission Systems Utilizing Optical Solitons — Results of Computer Simulation

In a conventional optical transmission system with transmission rate of 1 Gbit/s, the deterioration of the pulse shape is primarily due to fiber loss and, hence, the transmission distance is limited by the number of photons at the output side, so that the detector can operate electronically. Normally, this distance is about 100 kms. However, when the transmission speed is increased beyond 1 Gbit/s, the pulse distortion is determined by the group velocity dispersion and the transmission distance decreases in inverse proportion to the square of the transmission speed. In order to minimize the pulse distortion due to the dispersion, a dispersion shifted fiber may be used. However, the choice of zero group dispersion usually increases the fiber loss and, furthermore, makes it difficult to design a frequency multiplex transmission system.

A conceivable optical soliton transmission system utilizes periodic amplification of the soliton by means of the stimulated Raman process [6.7, 17] to reshape the pulse width, which otherwise tends to spread due to fiber loss. Because the band width of the Raman gain is very wide ($\cong$ 10 THz), this system allows one to design a frequency multiplex system with more than 10 channels, which makes it possible to design a super-fast transmission system of more than 100 Gbits/s.

A conventional linear transmission system requires repeaters to reshape the pulse shape. A repeater consists of a detector, a laser and a modulator. Becuase of this, the bit rate of a transmission system utilizing a linear pulse train is limited by the electronic response time of the detectors and modulators. For this reason, the cost of high bitrate optical transmission systems is determined by the cost of the repeaters.

However, in an optical soliton transmission system, only optical amplifiers which utilize the Raman effect in the same fiber are required and, therefore, the cost of the repeater is expected to be significantly lower.

The behaviour of the electric field of a soliton E_s under the influence of the electric field E_p of a Raman pump may be described by

$$\mathrm{i}\left[\frac{\partial E_p}{\partial x}+(\gamma_p+\alpha\,|E_s|^2)\,E_p-k_p'\,\frac{\partial E_p}{\partial t}\right]-\tfrac{1}{2}\,k_p''\,\frac{\partial^2 E_p}{\partial t^2}+g\,|E_p|^2E_p=0\,, \quad (6.1)$$

$$\mathrm{i}\left[\frac{\partial E_s}{\partial x}+(\gamma_s-\alpha\,|E_p|^2)\,E_s-k_s'\,\frac{\partial E_s}{\partial t}\right]-\tfrac{1}{2}\,k_s''\,\frac{\partial^2 E_s}{\partial t^2}+g\,|E_s|^2E_s=0\ . \quad (6.2)$$

Here, x is the transmission distance, t is the time, γ is the linear loss rate, k is the wave number, and the primes are derivatives with respect to the frequency ω. The subscripts p and s designate the pump and soliton quantities, respectively. The electric field E_p of the pump is assumed to have a band width large enough ($\gtrsim$15 GHz) to suppress the stimulated Brillouin scattering. The coupling coefficient α of the Raman process is related to the coefficient of the self-phase modulation g through the imaginary and real parts of the third-order susceptibility $\chi^{(2)}$ of silica [6.18],

$$\alpha \simeq 0.2\,g \quad .$$

As an example of an optical soliton system, let us consider [6.19] a fiber with an effective cross-sectional area of 20 μm^2 and a loss rate of 0.2 dB/km. We consider a soliton with a carrier wavelength of 1.55 μm, an amplitude of 30 mW, a pulse width of 10.2 ps and a repetition cycle of 100 ps (10 Gbit/s). For the purpose of reshaping the soliton, we consider a Raman pump which is injected in the forward and backward directions of the fiber at every 34.4 kms, having a cw (continuous wave) power of 40 mW and a wavelenght at 1.45 μm. By injecting pump waves in both directions, one can maintain a relatively constant gain along the fiber, in spite of the fiber loss which tends to dissipate the pump intensity. To decrease the required pump intensity we assume here a fiber with polarization reservation so that the Raman gain can be kept relatively high by choosing the polarization of the pump and the soliton in the same direction. As far as practical application is concerned, this may not be necessary if the pump has enough intensity. We note here that the Kerr effect which is needed for the formation of a soliton does not depend on the direction of polarization, but the Raman gain depends on the direction of the angle of polarization between the pump wave and the Stokes wave (soliton) and the gain differs by about a factor of 5 between the cases of parallel and perpendicular polarization.

When the pump wave is injected periodically along the fiber, the Raman gain also changes periodically. Because of this, the pulse width of a soliton shrinks in the portion where the Raman gain exceeds the fiber loss, while it expands in the region where the Raman gain is smaller than the loss rate. Consequently, two solitons which propagate side by side can interact with each other and form a bound oscillation [6.19–25]. In order to avoid this interaction, the separation between two solitons shoud be made suffciently large ($\gtrsim$8 times the soliton width) for a given distance of pumped power injection.

Figure 6.2 shows the result of computer simulation, using (6.1) and (6.2), for the transmission of a soliton pair with 100 ps separation, amplified periodically

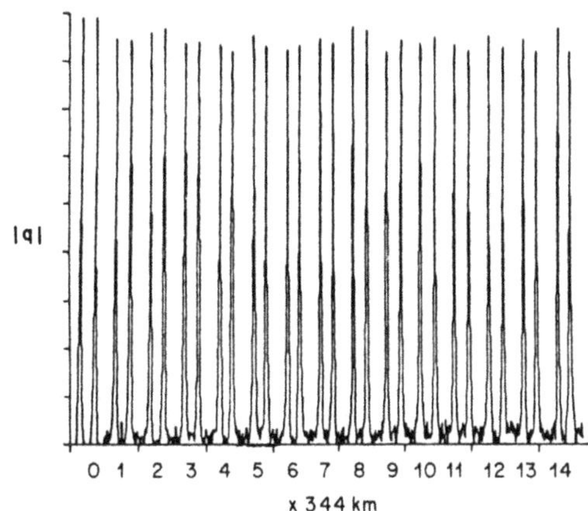

Fig. 6.2. Results of computer simulation of a pair of optical soliton transmissions periodically amplified by Raman processes. Outputs at every 344 km of a pair of 10 ps solitons with 100 ps spacing are shown [6.13]

by the Raman process [6.19]. The shape of the soliton pulse pair is taken at each 100th Raman amplification point. A 1 dB loss is assumed at each point where the Raman pump is injected. From this figure, one can see that the soliton pair can propagate stably for a distance of 4816 km.

The problem of noise accumulation in the fiber when Raman amplification is repeatedly applied to the system is not found to be serious in the case of transmission over several thousand kilometers since the photon number in one soliton is relatively large [6.26].

6.3 Experiment on the Long Distance Transmission of a Soliton by Repeated Raman Amplifications

In 1988, Mollenauer and his group made another break-through in soliton experiments. *Mollenauer* and *Smith* [6.27] succeeded in transmitting a soliton over 4.000 kms in a fiber with loss periodically compensated for by Raman gain, and this distance was soon extended to over 6000 kms [6.28].

The apparatus is shown schematically in Fig. 6.3. A 41.7 km length of low-loss (0.22 dB/km at 1600 nm) single-mode fiber with group-delay dispersion D of -17 ps/(nm km) at 1600 nm is closed by means of an all-fiber version of a Mach–Zehnder interferometer. The interferometer allows pump light at $\sim$ 1497 nm ($\sim$ 300 mW cw from a KCl : Tl color-center laser) to be efficiently coupled into the loop, while at the same time allowing about 95 % of the signal light (55 ps, minimum band width pulses from a mode-locked, 100 MHz pulse repetition rate, NaCl color-center laser operating at $\sim$ 1600 nm) to recirculate around the loop. The difference in the pump and signal frequencies, $\sim$ 430 cm^{-1}, corresponds to the peak of the Raman gain band in quartz glass. The 55 ps pulse width makes the soliton period $z_0 \sim 66$ km, so that we easily meet the criterion $z_0 \gg L/8$ for stable soliton transmission [6.12]. (L is the amplification period, here 41.7 kms.) The 5 % sample of the signal train leaving the loop is detected by an ultrafast diode (response time 9 ps), whose output is sent to a microwave spectrum analyzer. The pulse shape is inferred from the measured pulse envelope spectrum.

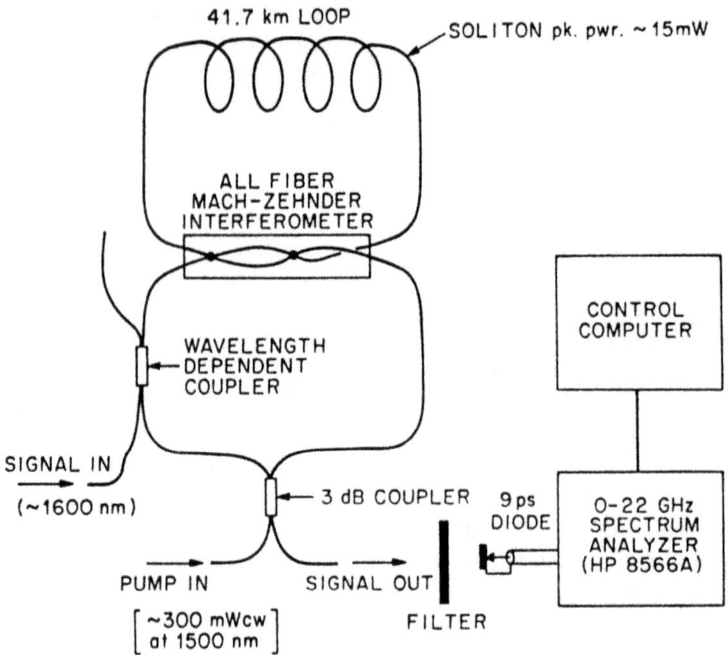

Fig. 6.3. Schematic diagram of the experiment of *Mollenauer* and *Smith* [6.22] of optical soliton transmission beyond 4000 kms by repeated Raman amplifications

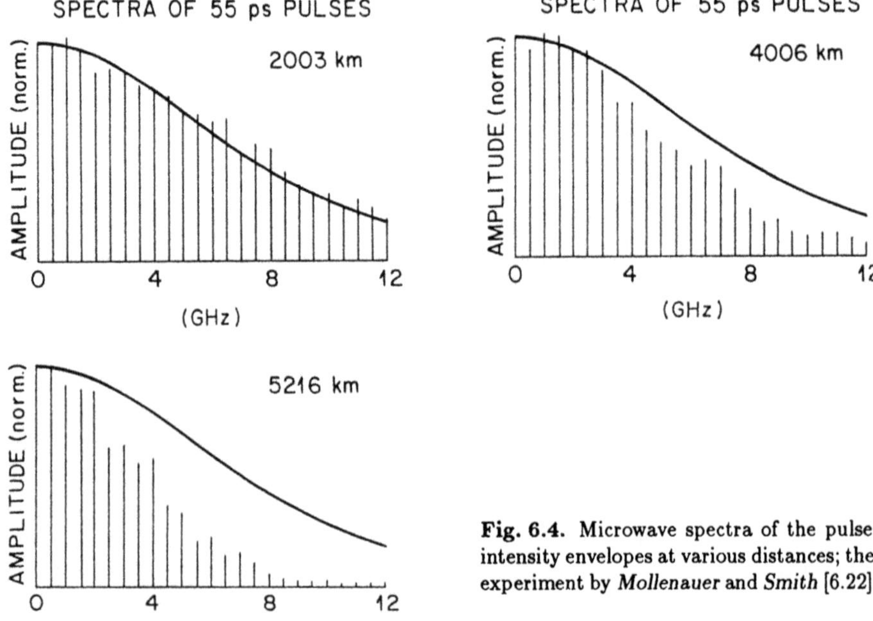

Fig. 6.4. Microwave spectra of the pulse intensity envelopes at various distances; the experiment by *Mollenauer* and *Smith* [6.22]

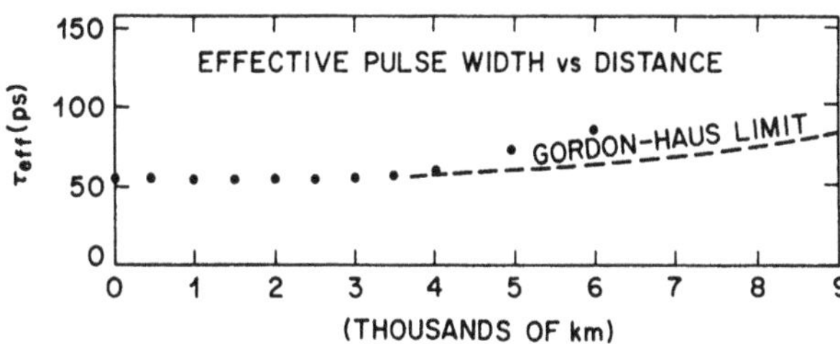

Fig. 6.5. Variation of the soliton pulse width in distance of transmission as obtained by *Mollenauer* and *Smith* [6.23]

In order to avoid stimulated Brillouin backscattering of the pump light, the pump laser's output was spread over 50 or more discrete frequencies in a band at least several gigahertz wide.

Figure 6.4 shows the microwave spectra of the pulse intensity envelopes for 48, 96, and 125 round trips (2003, 4006, and 5216 kms) and for the peak signal power in the fiber at the soliton value [6.27]. This result demonstrates that there is little variation in the pulse width in transmission over these distances.

Figure 6.5 shows the variation in the effective transmitted pulse width of a soliton over the transmission distance most recently obtained by *Mollenauer* and *Smith* [6.28]. The result clearly shows that the pulse width remained narrow beyond 6.000 kms, approaching the theoretical limit obtained by *Gordon* and *Haus* [6.26]. These experimental results demonstrate that optical solitons have a possible practical application in a high bit rate all-optical communication system. In particular, the wide band gain of the Raman amplification provides for the possibility of a large capacity, multi-channel soliton transmission system [6.29] costing considerably less than a conventional system which requires repeated regeneration of the pulse train and, therefore, involves repeaters containing a complicated combination of electronics and optics.

The Raman amplification also has applications in the generation [6.30] and compression [6.21, 32] of solitons. These will be discussed in the next chapter.

6.4 Amplification by Erbium Doped Fiber

In addition to its ability as a Raman amplifier, a fiber can also act as a laser amplifier when it is doped with a rare earth metal [6.12–16]. It can thus be used to reshape solitons. For example, *Nakazawa* et al. [6.16] have recently succeeded in amplifying a 20 Gbit/s burst pulse train of solitons by using a 3.5 m erbium doped fiber. Figure 6.6 shows the results of the experiments by *Nakazawa* et al. [6.16]. Fiber A has a length of 27 km and a dispersion of $-3.8\,\mathrm{ps(nm\,km)}^{-1}$ at a signal wavelength of $1.552\,\mu$m. The theoretical $N = 1$ soliton peak power is about 50 mW for a pulse width of 10 ps and the soliton period is approximately 10 km. For an erbium pump power of 23 mW, the output pulse train is overlapped by the linear dispersion, as seen in Fig. 6.6a. As the pump power is increased to 45 mW and to 90 mW, the separation of each pulse is clearly seen (Fig. 6.6b–d).

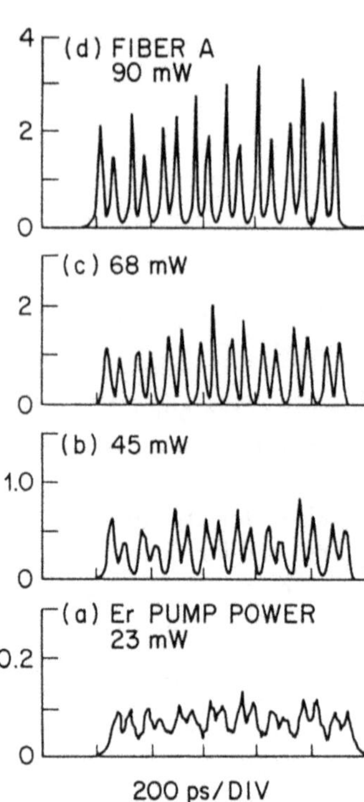

Fig. 6.6a–d. Experimental results of *Nakazawa* et al. [6.16]. As the Er pump power is increased, the pulse train is seen to separate forming solitons

One advantage of this technique is that it requires relatively low pump powers in comparison to a Raman amplifier. However, due to the gain saturation for a large signal power level, amplification of a continuous train of solitons beyond 10 Gbit/s requires the amplifiers to be spaced apart at approximately 10 km or less. This spacing is desirable for a stable transmission of solitons over several thousand kilometers when periodic localized amplifiers are used [6.2].

The erbium doped fiber amplifier can be used as an important element of an all-optical soliton transmission system utilizing only laser diodes. The erbium doped fiber amplifier can amplify a small amplitude modulated pulse train generated by a laser diode to sufficiently large amplitudes to form solitons. Then the modulated soliton train can be transmitted through a fiber without distortion by repeated Raman amplifications. Laser diodes can be used for the pumping of the erbium doped fiber as well as for the Raman pumping of the transmission line fiber. Such an experiment was recently performed by *Iwatsuki* et al. [6.33]. An optical pulse train at a 2.8 GHz repetition rate is generated by a 1.55 μm distributed feedback laser diode and modulated by NRZ (Non Return to Zero) pseudo random pattern using a Ti : $LiNbO_3$ Mach-Zehnder intensity modulator. Fabry-Perot laser diodes are used as pump sources to amplify the optical pulses to form solitons as well as to Raman-compensate the fiber loss. Figure 6.7 shows the optical signal pulse modulated by $\{1.0\}$ fix pattern after transmitting through the dispersion shifted fiber with a length of 23 km, D of $-4.0\,\mathrm{ps(nm\,km)}^{-1}$ a loss of 0.25 dB/km, and a mode-field diameter of 6.0 μm at 1.55 μm signal wavelength.

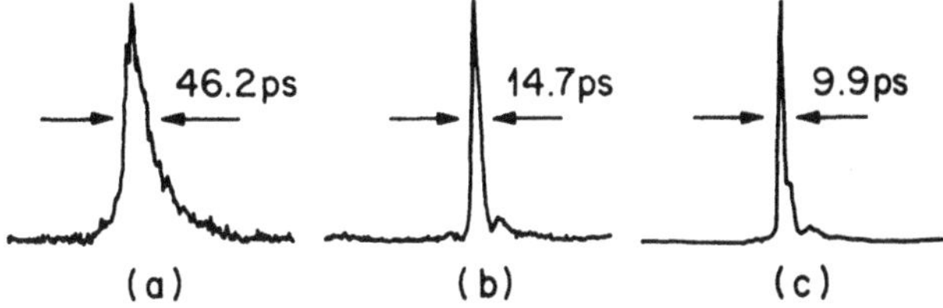

Fig. 6.7a–c. Experimental results of *Iwatsuki* et al. [6.33]. Reduction of the output pulse width is clearly seen for increased input peak power from 9.3 mW (**a**) to 155 mW (**b**) and to 540 mW (**c**)

The total compensation by the Raman pump for the loss of the fiber is 0.5 dB. The pulse narrowing and compression are clearly seen in the figure as the input peak power is raised from 9.3 mW, which is subcritical to the theoretical soliton peak power of 18 mW, to beyond the critical power of 155 mW and 540 mW.

7. Other Applications of Optical Solitons

In addition to its application in optical communication systems, the soliton phenomenon can be used for many other purposes. In this chapter we give some examples of the uses of the soliton phenomenon.

7.1 Soliton Laser

An optical soliton can be produced in a fiber by injecting a mode-locked laser pulse with an appropriate amplitude. However, by employing the fiber as part of the feedback system in the cavity of the laser oscillator, one can produce a soliton with arbitrary width. The soliton laser invented by *Mollenauer* et al. [7.1] utilizes this mechanism. It succeeded in producing a subpicosecond pulse at infra-red wavelength for the first time.

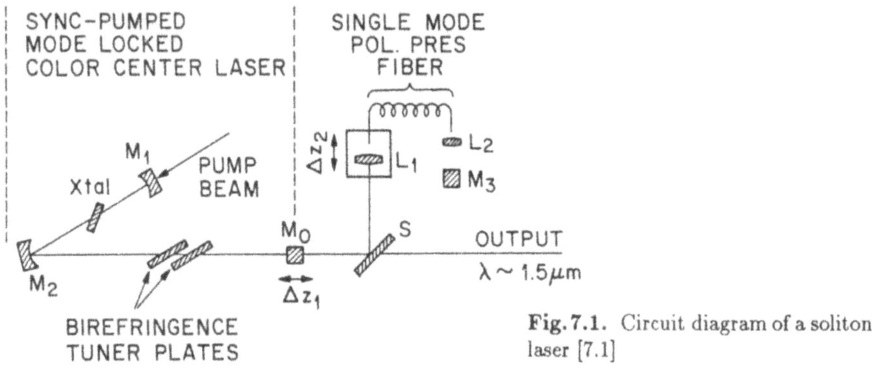

Fig. 7.1. Circuit diagram of a soliton laser [7.1]

Mollenauer et al. [7.1] constructed $N = 2$ solitons by utilizing the circuit shown in Fig. 7.1. In this figure, the output signal of the mode-locked laser is inserted into the fiber with anomalous dispersion through the beam splitter S. The light which is reflected is fed back through the same fiber, and inserted back into the laser cavity. The pulse width of the laser output pulse shrinks as it propagates back and forth in this feedback loop and can be extracted as a soliton. For the successful operation of this system, a polarization preserving fiber should be used. In addition, a variable path portion is required at the input terminal, so that the optical distance of the fiber can be adjusted to the integer multiple of the distance of the main cavity.

One somewhat puzzling property of the soliton laser is that the output soliton seems to have a stable operation only at $N = 2$ solitons. *Haus* and *Islam* [7.2] gave a theoretical interpretation of this. Theories of the soliton laser have also been developed by several other authors [7.3, 4], and its stability has been discussed experimentally [7.5] as well as theoretically [7.6].

The Raman gain in fibers can be used to construct a laser known as the Raman fiber laser [7.7–9]. Soliton lasers have also been constructed using Raman amplification in fibers [7.10, 11].

7.2 Soliton Compression

When the intensity of the input signal is larger than the intensity of a soliton for the given pulse width, more than one soliton is formed. If the input pulse has a symmetric shape, the solitons generated propagate at an identical speed, in the absence of higher order effects [7.12]. Because of this, phase interference occurs among the solitons and, at a particular point during transmission, the pulse width is compressed [7.13–18]. If one utilizes this fact, pulse compression is possible. Figure 7.2 is an example of such a pulse compression experiment. Here, a 150 fs pulse is compressed to 50 fs by transmitting it through a 35 cm fiber [7.13].

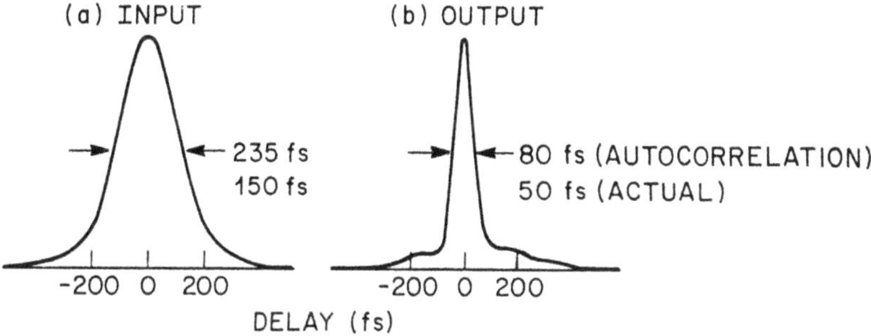

Fig. 7.2. Experimental result of soliton compression [7.13]

Pulse compression can also be achieved in the normal dispersion regime ($k'' > 0$) by reversing the group dispersion, using a grading pair [7.19]. If one uses this principle in combination with the soliton phenomenon, a compression with an extremely large factor can be achieved. The Nd : YAG laser has a wavelength of 1.32 μm, which is close to the zero group dispersion wavelength of a dielectric material. By a proper choice of the profile of the refractive index in the cross-section of a fiber, one can construct a dispersion shifted fiber, i.e., the fiber can have either an anomalous dispersion ($k'' < 0$), or a normal dispersion ($k'' > 0$) at this wavelength. By utilizing two types of fiber, the pulse can be compressed first in the normal dispersion fiber with the help of the grating pair, and then in the anomalous dispersion fiber with the help of the soliton compression mechanism. By this method, compression by a factor of 1000 to 3000 has been achieved experimentally [7.20–22]. However, as will be shown in Sect. 8.2, the self-induced

Raman effect in a fiber downshifts the frequency spectra of a soliton and leads to the splitting of the higher order solitons during compression [7.12, 23, 24]. In order to avoid this problem, it is desirable to use a Raman amplification so that a soliton is compressed adiabatically by keeping the $N = 1$ property of the soliton [7.25].

As with other applications of solitons, the use of solitons for switching has been considered by several authors [7.26–28]. The soliton switch, together with the application of modulational instability which will be discussed in Chap. 9, provides the possibility for a unique all-optical switching device.

8. Influence of Higher Order Terms

In this chapter we discuss the influence of the higher order terms in the nonlinear Schrödinger equation on soliton propagation in a fiber. In particular, we consider the effect of the self-induced Raman effect on soliton transmission.

The fact that the nonlinear Schrödinger equation can describe nonlinear pulse transmission in a fiber has been demonstrated by various experiments. Recently, effects which can be described only by taking into account the higher order terms in the nonlinear Schrödinger equation have been observed.

As was shown in Chap.4, the small parameter ε which is used in the derivation of the nonlinear Schrödinger equation has a size of the order of $(\omega\tau_0)^{-1}$, where τ_0 is the pulse width and ω the carrier frequency of the light wave. For 10 ps, for example, ε is of the order of 10^{-4} and consequently, the higher order term gives corrections of the order of 10^{-8}. The fact that such a small effect can be detected experimentally originates from the fact that the laser frequency is almost monochromatic at 10^{14} Hz. Consequently, the effect of a 10^{-8} correction represents a correction in the frequency of the order of 10^6 Hz and can, therefore be detected relatively easily.

As was shown in Sect. 3.4, the nonlinear Schrödinger equation is modified by introducing higher order terms. The modified nonlinear Schrödinger equation may be expressed in the following form [8.1]:

$$\mathrm{i}\frac{\partial q}{\partial Z} + \frac{1}{2}\frac{\partial^2 q}{\partial T^2} + |q|^2 q + h = 0 \quad , \tag{8.1}$$

where h represents the higher order effects,

$$h = \mathrm{i}\varepsilon\left\{\beta_1 \frac{\partial^3 q}{\partial T^3} + \beta_2 \frac{\partial}{\partial T}\left(|q|^2 q\right) + \mathrm{i}\sigma_3 q\, |q|^2\right\} \quad . \tag{8.2}$$

In (8.2), the first term represents the linear higher order dispersion effect, (5.15),

$$\beta_1 = \frac{1}{6}\frac{k'''\lambda}{T_0^3} \quad , \qquad \lambda = \frac{2\pi}{k} \quad . \tag{5.15}$$

The second term represents the nonlinear dispersion effect which originates from the wavelength dependency of the Kerr coefficient n_2, and is given by (5.16),

$$\beta_2 = \frac{1}{gT_0}\frac{\partial}{\partial\omega}\left[\frac{\omega^2}{k\,c^2 S_0 E_0^4}\int n_0 n_2\, |\nabla\phi|^4\, dS\right] \quad . \tag{5.16}$$

The last term represents the self-induced Raman effect which produces the down shift of the soliton spectrum by the Raman induced spectral decay. The coefficient of this term is given by

$$\sigma_3 = -\mathrm{i}\,\beta_3 \quad , \tag{8.3}$$

with β_3 given by (5.17),

$$\beta_3 = \frac{1}{gT_0}\,\frac{\omega^2}{kc^2 S_0 E_0^4} \int \left[n_0 n_2 \,\frac{\partial\,|\nabla\phi|^4}{\partial\omega_0} + \frac{3}{4}\,(\chi_1^{(2)} - \chi_{-1}^{(2)})\,|\nabla\phi|^4 \right] dS \quad . \tag{5.17}$$

Of these three higher order terms, the last term which represents the self-induced Raman effect plays the most dominant role, as will be shown in Sect. 8.1 and 8.2. The first two terms which can be derived by a phenomenological expansion of the nonlinear refractive index (shown in Sect. 3.2), have been included in the analyses presented by several authors [8.2–7]. The nonlinear Schrödinger equation including the first two terms is still integrable [8.1] in this order and the soliton property is not essentially modified, as will be shown in Sect. 8.3.

The effect of these higher order terms on the transmission property of solitons can be obtained by constructing the conservation laws shown in Sect. 4.3,

$$\mathrm{i}\,\frac{\partial}{\partial Z} \int_{-\infty}^{\infty} |q|^2\, dT = - \int_{-\infty}^{\infty} (hq^* - h^*q)\, dT = 0 \tag{8.4}$$

$$\begin{aligned} &\frac{\mathrm{i}}{2}\,\frac{\partial}{\partial Z} \int_{-\infty}^{\infty} \left(q\,\frac{\partial q^*}{\partial T} - q^*\,\frac{\partial q}{\partial T} \right) dT = - \int_{-\infty}^{\infty} \left(h\,\frac{\partial q^*}{\partial T} + h^*\,\frac{\partial q}{\partial T} \right) dT \\ &= \varepsilon\sigma_3 \int_{-\infty}^{\infty} \left(\frac{\partial}{\partial T}\,|q|^2 \right)^2 dT - \mathrm{i}\,\varepsilon\beta_2 \int_{-\infty}^{\infty} \left(q\,\frac{\partial q^*}{\partial T} - q^*\,\frac{\partial q}{\partial T} \right) \frac{\partial}{\partial T}\,|q|^2\, dT \quad . \end{aligned} \tag{8.5}$$

In particular, we note that these higher order terms do not change the soliton energy, as is seen from (8.4). However, the momentum is modified, as shown in (8.5).

8.1 Self-Frequency Shift of a Soliton Produced by Induced Raman Scattering

A phenomenon which cannot be described by the ideal nonlinear Schrödinger equation was first observed by *Mitschke* and *Mollenauer* [8.8], who detected the shift of the central frequency of a soliton to a lower frequency when the soliton width was less than 1 ps. *Gordon* [8.9] has successfully interpreted the result in terms of the stimulated Raman process, in which the central frequency of the soliton spectrum amplifies the lower sideband frequency components.

Kodama and *Hasegawa* [8.1] have identified this effect in terms of the third term in (8.2) and obtained a result identical to that of Gordon.

If we take the effect of the self-induced Raman term (the third term) and use one-soliton solution of (3.31) in q of (8.4) and (8.5), we have

$$\frac{d\eta}{dZ} = 0 \tag{8.6}$$

$$\frac{d\kappa}{dz} = -\frac{8}{15}\,\varepsilon\,\sigma_3\,\eta^4 \quad . \tag{8.7}$$

Here, η is the normalized amplitude of the soliton, and κ represents the frequency of the soliton. Equation (8.7) shows that the soliton frequency decreases in proportion to the fourth power of its amplitude. If we use the original parameters, (8.7) reduces to

$$\frac{df}{dx} = \frac{4}{15}\,\frac{n_2}{\lambda k''}\,\beta\,E_0^4 = \frac{2.56}{\pi^2}\,\frac{\lambda k''\beta}{n_2\tau_0^4} \quad . \tag{8.8}$$

Here, τ_0 is the width of the soliton and use is made of (5.20). In this expression, the coefficient β can be expressed in terms of the differential gain, γ_{R}, with respect to the frequency separation $\Delta\omega$ between the pump and the Stokes frequencies,

$$\beta = \left|\frac{\partial\gamma_{\mathrm{R}}}{\partial\Delta\omega}\right| \quad , \tag{8.9}$$

where $\gamma_{\mathrm{R}}E_0^2$ gives the Raman gain per unit length of the fiber.

8.2 Fission of Solitons Produced by Self-Induced Raman Scattering

The fact that the central frequency of a soliton decreases in proportion to the distance of propagation implies that the group velocity decreases by the factor $\Delta v_{\mathrm{g}} = (\partial v_{\mathrm{g}}/\partial\omega)\Delta\omega = -k''\Delta\omega/(k')^2\,(< 0)$. Since $\Delta\omega$ is proportional to the fourth power of the soliton amplitude, the decrease in the soliton speed becomes larger for a soliton with larger amplitude.

As was discussed in Chap.4, in the absence of the self-induced Raman process, solitons formed with the initial amplitude $A \geq N$ $(N \geq 2)$ propagate at the same speed, with phase interference. However, in the presence of the self-induced Raman process, these N number of solitons propagate at different speeds and hence, they separate [8.10–12]. Figure 8.1 is the numerical result [8.12], with the boundary condition that at $Z = 0$, $q(T,0) = 3\operatorname{sech} T$, which describes the behaviour of solitons in the presence of the self-induced Raman effect. The numerical calculation assumes a periodic boundary condition with a period of $T = 50$, and the value of $\varepsilon\sigma_3$ in expression (8.2) was taken to be 10^{-3}.

Figure 8.1a represents the magnitude of q, $|q|$, at $Z = 20$. Figure 8.1b represents the contour of $|q|$ between $Z = 0$ to 20. The amplitudes of the 3 solitons which, from the inverse scattering calculation, (4.12), are expected to be produced from this initial condition are 5, 3 and 1. As is seen from Fig. 8.1b, the soliton with $\eta = 5$ is first ejected. At this moment, due to the approximate con-

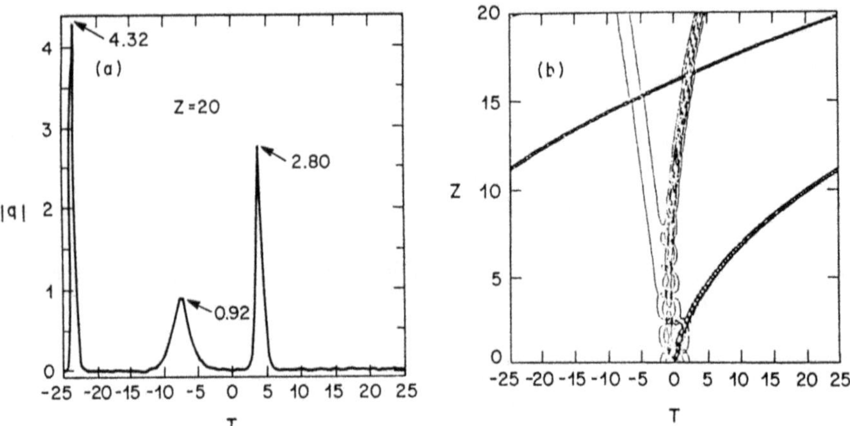

Fig. 8.1a,b. Numerical result of fission of $N = 3$ solitons produced by self-induced Raman process [8.11]

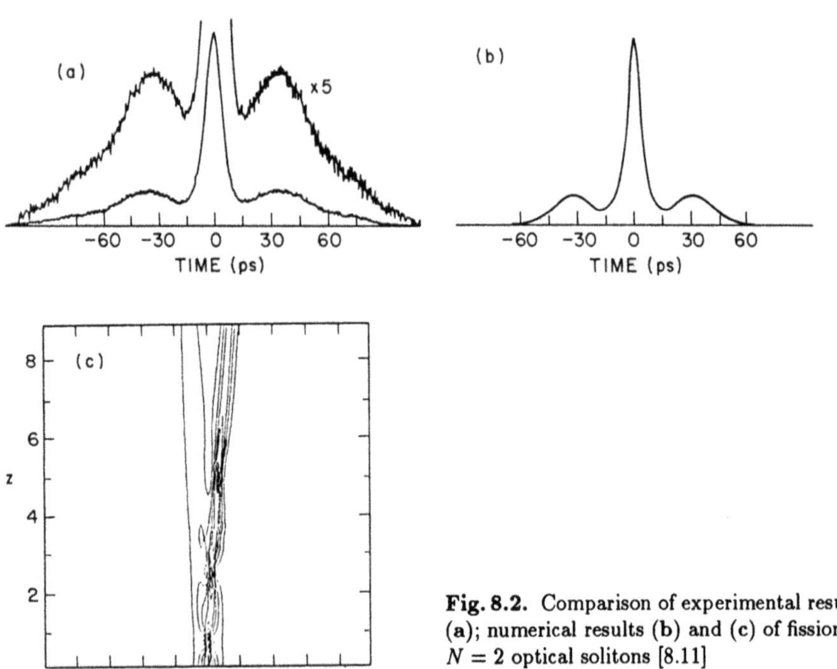

Fig. 8.2. Comparison of experimental results (**a**); numerical results (**b**) and (**c**) of fission of $N = 2$ optical solitons [8.11]

servation of the momentum, the other solitons shift slightly to the left. Later, after a few oscillations, the solitons with $\eta = 3$ and $\eta = 1$ are found to split. From Fig. 8.1a, the ratio of the amplitudes of the three solutions can be identified as $1 : 3 : 5$, as is theoretically predicted from the inverse scattering calculation.

The experimental verification of the fission of solitons has been observed by several authors [8.12–14]. Figure 8.2 shows the comparison between the experimentally obtained autocorrelation spectrum of two solitons which are produced from the initial condition 2 sech T and that obtained numerically, where the induced Raman term of (8.2) is included [8.12].

8.3 Effects of Other Higher Order Terms

In the absence of the Raman term, σ_3 , (8.1) was shown by *Kodama* [8.15, 16] to be integrable. First, Kodama noted that (5.6) can be transformed into a higher-order nonlinear Schrödinger equation given by [8.15, 16]

$$\mathrm{i}\,\frac{\partial Q}{\partial Z}+\frac{1}{2}\,\frac{\partial^2 Q}{\partial T^2}+|Q|^2\,Q+\varepsilon\,\mathrm{i}\,\beta_1\left(\frac{\partial^3 Q}{\partial T^3}+6\,|Q|^2\,\frac{\partial Q}{\partial T}\right)=O(\varepsilon^2)\quad . \tag{8.10}$$

Equation (8.10) is also found to be completely integrable by means of the inverse scattering transform having the same eigenvalue problem as (4.6) an (4.7)

$$\mathrm{i}\,\frac{\partial \psi_1}{\partial T}+q_0(T)\,\psi_2=\zeta\,\psi_1 \tag{4.6}$$

$$-\mathrm{i}\,\frac{\partial \psi_2}{\partial T}-q_0^*(T)\,\psi_1=\zeta\psi_2\quad . \tag{4.7}$$

Here, the transformation $\Gamma_\varepsilon : q \rightarrow Q$ is given by

$$\begin{aligned} Q=\Gamma_\varepsilon(q)=&q-\varepsilon\,\mathrm{i}\left(3\beta_1-\tfrac{1}{2}\,\beta_2\right)\frac{\partial q}{\partial T}\\ &-\varepsilon\,\mathrm{i}(6\beta_1-2\beta_2-\beta_3)\,q\int_{-\infty}^{T}|q(T')|^2\,dT'+O(\varepsilon^2)\quad . \end{aligned} \tag{8.11}$$

Using this transform and the one-soliton solution (4.9) of (4.6) and (4.7), one can obtain the one-soliton solution of (8.10),

$$Q(T,Z)=\eta\,\mathrm{sech}\,\{\eta(T-vZ-\theta_0)\}\cdot\exp(-\mathrm{i}\kappa T+\mathrm{i}\omega Z-\mathrm{i}\sigma_0)\quad , \tag{8.12}$$

where the velocity and frequency of the soliton are given by $v=-\kappa+\varepsilon\beta_1(\eta^2-3\kappa^2)$ and $\omega=(\eta^2-\kappa^2)/2+\varepsilon\beta_1\kappa(3\eta^2-\kappa^2)$, respectively. It should be noted that this solution results from the same initial condition as the nonlinear Schrödinger soliton (4.9), and the velocity of the soliton is modified by an amount $\Delta v=\varepsilon\beta_1(\eta^2-3\kappa^2)$ due to the higher-order dispersion. Because of this velocity deviation Δv which depends on the soliton parameters, the bound N-soliton solutions of the nonlinear Schrödinger equation split into individual solitons. The splitting of solitons caused by these two terms (involving β_1 and β_2) , however, is less manifest than the Raman term (involving $\mathrm{Im}\{\beta_3\}$) since the Raman term gives a constant deceleration and, hence, the deviation in the velocity increases in proportion to the distance of propagation.

9. Modulational Instability

A continuous wave in the anomalous dispersion regime is known to produce modulational instability. Modulation in frequency and amplitude grows exponentially. This process is discussed in this chapter.

9.1 Modulational Instability in the Nonlinear Schrödinger Equation

As has been shown, the light wave in a fiber can be described by the nonlinear Schrödinger equation. When the input wave has a pulse shape, the output can be described in terms of a set of solitons and a dispersive wave as shown from the inverse scattering calculation. The question we should like to discuss in this chapter is, what happens if the input light wave has a continuous amplitude?

Let us start with the description of the wave envelope q given by (5.2),

$$\mathrm{i}\frac{\partial q}{\partial Z}+\frac{1}{2}\frac{\partial^2 q}{\partial T^2}+|q|^2 q=-\mathrm{i}\,\Gamma q \quad . \tag{9.1}$$

Here, we show that the input light wave becomes unstable for a small perturbation around the initial amplitude q_0. This instability is called the modulational instability. To show the instability, we introduce new real variables ϱ and σ through

$$q=\sqrt{\varrho}\,\mathrm{e}^{\mathrm{i}\sigma} \tag{9.2}$$

and substitute (9.2) into the nonlinear Schrödinger equation (9.1) to obtain

$$\frac{\partial \varrho}{\partial Z}+\frac{\partial \varrho}{\partial T}\frac{\partial \sigma}{\partial T}+\varrho\frac{\partial^2 \sigma}{\partial T^2}+2\,\Gamma\varrho=0 \tag{9.3}$$

and

$$\varrho-\frac{\partial \sigma}{\partial Z}+\frac{1}{4\varrho}\frac{\partial^2 \varrho}{\partial T^2}-\frac{1}{2}\left(\frac{\partial \sigma}{\partial T}\right)^2-\frac{1}{8\varrho^2}\left(\frac{\partial \varrho}{\partial T}\right)^2=0 \quad . \tag{9.4}$$

We consider a small modulation of ϱ and σ with the side band frequency given by Ω, such that

$$\varrho(T,Z)=\varrho_0(Z)+\mathrm{Re}\left\{\varrho_1(Z)\,\mathrm{e}^{-\mathrm{i}\Omega T}\right\} \tag{9.5}$$

and

$$\sigma(T,Z) = \sigma_0(Z) + \mathrm{Re}\left\{\sigma_1(Z)^{-\mathrm{i}\Omega T}\right\} \quad . \tag{9.6}$$

If we substitute (9.5) and (9.6) into (9.3) and (9.4) and linearize the results, from the zeroth order terms, we have

$$\varrho_0 - \frac{\partial \sigma_0}{\partial Z} = 0 \tag{9.7}$$

$$\frac{\partial \varrho_0}{\partial Z} + 2\Gamma \varrho_0 = 0 \quad . \tag{9.8}$$

Equations (9.7) and (9.8) are easily solved for the initial condition $\varrho_0(0) = \overline{\varrho}_0$ and $\sigma_0(0) = 0$, giving

$$\varrho_0(Z) = \overline{\varrho}_0\, \mathrm{e}^{-2\Gamma Z} \tag{9.9}$$

$$\sigma_0(Z) = \frac{\overline{\varrho}_0}{2\Gamma}\left(1 - \mathrm{e}^{-2\Gamma Z}\right) \quad . \tag{9.10}$$

Equation (9.9) simply indicates that the average (carrier) intensity decreases exponentially at the rate of 2Γ, where $\overline{\varrho}_0$ is the initial intensity of the carrier. From the first order terms, we have equations for the side band amplitude and phase,

$$\frac{d\varrho_1}{dZ} + 2\Gamma \varrho_1 - \Omega^2 \varrho_0 \sigma_1 = 0 \tag{9.11}$$

and

$$\varrho_1\left(1 - \frac{\Omega^2}{4\varrho_0}\right) - \frac{d\sigma_1}{dZ} = 0 \quad . \tag{9.12}$$

In order to illustrate the instability, let us ignore the fiber loss Γ here and write $\varrho_1(Z)$ and $\sigma_1(Z)$ in terms of the Fourier amplitude, $\varrho_1(Z) = \mathrm{Re}\left\{\varrho_1\, \mathrm{e}^{\mathrm{i}kZ}\right\}$ and $\sigma_1(Z) = \mathrm{Re}\left\{\sigma_1\, \mathrm{e}^{\mathrm{i}kZ}\right\}$. Then, (9.11) and (9.12) give the following dispersion relation for the wave number K and frequency Ω:

$$K^2 = \tfrac{1}{4}\left(\Omega^2 - 2\varrho_0\right)^2 - \varrho_0^2 \quad . \tag{9.13}$$

This expression gives the spatial growth rate $\mathrm{Im}\{K\}$ which achieves its maximum value at

$$\Omega \equiv \Omega_{\mathrm{m}} = \sqrt{2\varrho_0} = \sqrt{2}\,|q_0| \quad , \tag{9.14}$$

and the corresponding growth rate becomes

$$\mathrm{Im}\, K = \varrho_0 = |q_0|^2 \quad . \tag{9.15}$$

If we write the variables in terms of the original parameters, the frequency that shows the maximum growth rate is given by

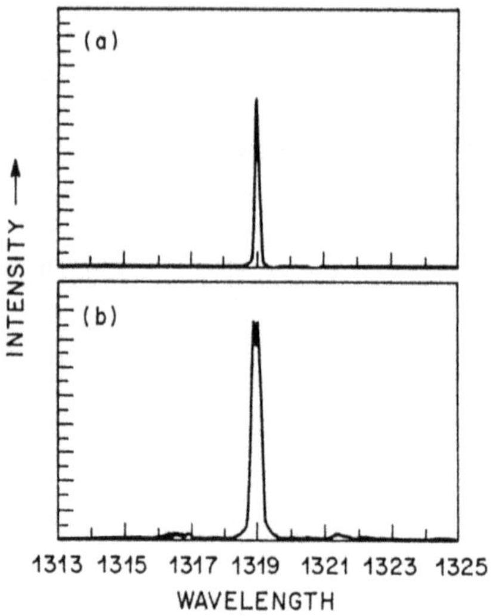

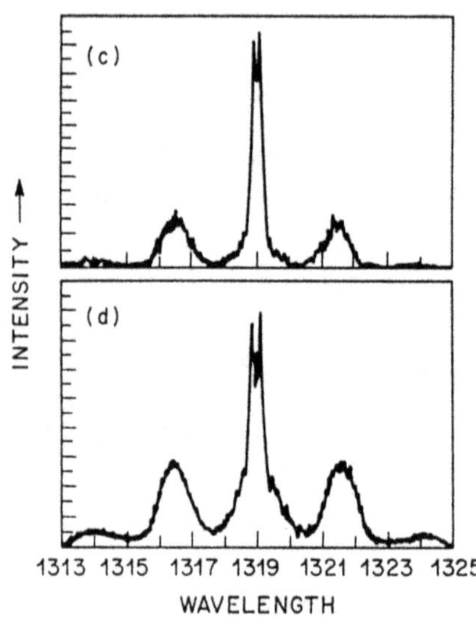

Fig. 9.1a–d. Experimental observation of modulational instability [9.8]. Input power level low (**a**); 5.5 W(**b**); 6.1 W(**c**); 7.1 W(**d**). For details see text

$$\omega_{\mathrm{m}} = \frac{\sqrt{2}\, E_0\, \pi\, n_2}{(-\lambda k'')^{1/2}} \quad , \tag{9.16}$$

and the corresponding spatial growth rate γ_{g} is given by

$$\gamma_{\mathrm{g}} = \pi\, n_2\, |E_0|^2/\lambda \quad . \tag{9.17}$$

If we recognize that ω_{m}^{-1} corresponds approximately to the pulse width of a soliton with amplitude E_0, we can see a close relationship between the formation of a soliton and modulational instability. The modulational instability of light waves in a fiber has been suggested by several theorists [9.1–7] as a means of producing a tunable light source [9.1], as a mechanism for the deterioration of coherent transmission capacity [9.2–4] and a source of cross phase modulation [9.6, 7].

Figure 9.1 shows the experimental result of the modulational instability of a light wave in a fiber [9.8]. The experiment is performed by injecting the pulses of a mode-locked Nd: YAG laser with a wavelength of 1.319 μm, a pulse width of 100 ps and a repetition rate of 100 MHz into a fiber. The 100 ps pulse is used instead of a constant amplitude wave in order to suppress the induced Brillouin scattering. The fiber used has a group dispersion of −2.4 ps/(nm km), a length of 1 km and an effective cross-section of 60 $\mu\mathrm{m}^2$, the loss rate at the wavelength 1.319 μm being 0.27 dB/km. Figure 9.1a shows the output power spectrum when the input power level is low. In this case, the output power spectrum corresponds to that of the input power. The finite width of the spectrum is a result of the 100 ps pulse width. When the input power is increased to 5.5 W (b), 6.1 W (c), 7.1 W (d), the side bands are found to be generated (b), then to grow (c), and higher order side bands are also found to be generated (d). In addition, we can recognize that the spacings between the side bands and the carrier frequency become wider as the intensity of the wave is increased. This is because, from (9.16), the side-band frequency ω_m, which gives the maximum growth, increases in proportion to $|E|^2$.

9.2 Effect of Fiber Loss

When the light intensity is low and the growth rate of the modulational instability is comparable to the fiber loss rate, one can not ignore the Γ term in (9.11) and (9.12). Let us now consider how the fiber loss modifies the modulational instability [9.2].

If we eliminate σ_1 from (9.11) and (9.12) and construct the differential equation for the normalized side band amplitude $\overline{\varrho}_1 = \varrho_1/\varrho_0$ (ϱ_0 is given by (9.9)), we get

$$\frac{d^2\overline{\varrho}}{dZ^2} - \Omega^2\left(\overline{\varrho}_0\,\mathrm{e}^{-2\Gamma Z} - \frac{\Omega^2}{4}\right)\overline{\varrho} = 0 \quad . \tag{9.18}$$

If we introduce a quantity R which designates the ratio of Ω^2 to ϱ_0, $R = \Omega^2/\overline{\varrho}_0$, R may be expressed in terms of engineering parameters as

$$R = \frac{\Omega^2}{\varrho_0} = 1.1 \times 10^4\,\frac{f^2 S}{P}\,(-\lambda^3 D) \quad , \tag{9.19}$$

where f (GHz) is the side band (modulation) frequency. P (mW) represents the light wave power, S (μm^2) the effective cross section of the fiber, λ (μm) the wavelength of the light and D (ps/(nm km)) the fiber dispersion. Equation (9.18) indicates that the side band with frequency f grows if $R<4$. For nominal parameters of $S = 60\ \mu\mathrm{m}^2$, $\lambda = 1.5\ \mu$m and $D = -10$ ps/(nm km), (9.19) with $R < 4$ yields for the relation between the critical unstable modulation frequency f_c (GHz) and the carrier power P (mW),

$$f_c < 4.2\sqrt{P} \quad . \tag{9.20}$$

If the fiber loss is taken into consideration, (9.18) indicates that the growth ceases at a distance Z_m, given by

$$Z_m = \frac{1}{2\Gamma}\ln\frac{4\overline{\varrho}_0}{\Omega^2} = -\frac{1}{2\Gamma}\ln\frac{R}{4} \quad . \tag{9.21}$$

Hence, the integrated exponential gain G over a distance $Z \geq Z_m$ is given by

$$G = \int_0^{Z_m} \mathrm{Im}\,\{K(Z)\,dZ\} \quad , \tag{9.22}$$

where

$$\mathrm{Im}\,\{K\} = \mathrm{Re}\left\{\Omega\left[\overline{\varrho}_0\,\mathrm{e}^{-2\Gamma Z} - \frac{\Omega^2}{4}\right]^{1/2}\right\} \quad . \tag{9.23}$$

G may be evaluated by introducing a new variable $u^2 = \mathrm{e}^{-2\Gamma Z} - \Omega^2/4\overline{\varrho}_0$ to give [9.2, 9]

$$G = \frac{\overline{\varrho}_0}{\Gamma}\sqrt{R}\left[\left(1-\frac{R}{4}\right)^{1/2} - \frac{\sqrt{R}}{2}\tan^{-1}\left(\frac{4}{R}-1\right)^{1/2}\right] \equiv \frac{\overline{\varrho}_0}{\Gamma} f(R) \quad , \qquad (9.24)$$

where R is defined in (9.19), and

$$\frac{\overline{\varrho}}{\Gamma} = 1.6\,\frac{P}{S\lambda\delta} \quad , \qquad (9.25)$$

δ being the fiber loss rate in dB/km.

For nominal parameters of $S = 60\ \mu\mathrm{m}^2$, $\lambda = 1.5\ \mu\mathrm{m}$ and $\delta = 0.2\ \mathrm{dB/km}$,

$$\frac{\overline{\varrho}_0}{\Gamma} = 8.7 \times 10^{-2} P(\mathrm{mW}) \quad , \qquad (9.26)$$

while $f(R)$ is plotted in Fig. 9.2 [9.11].

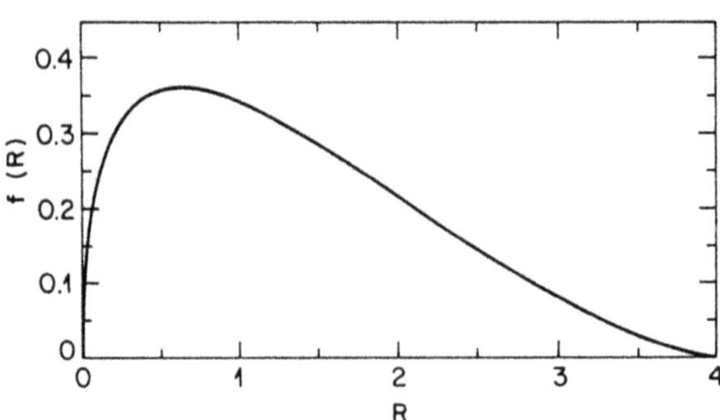

Fig. 9.2. $f(R)$ of (9.24)

We see that $f(R)$ has a maximum value of $\simeq 0.36$ at $R \simeq 0.6$. For $R \simeq 0.6$ and for $S = 60\ \mu\mathrm{m}^2$, $\lambda = 1.55\ \mu\mathrm{m}$, $D = -10$ ps/(nm km) and $\delta = 0.2$ dB/km (9.19) gives $f\ (\mathrm{GHz}) \simeq 1.6\sqrt{P(\mathrm{mW})}$, (9.24) gives $G \simeq 3.1 \times 10^{-2} P\ (\mathrm{mW})$, while (9.21) gives the un-normalized distance $Z_\mathrm{m} = 40$ km. It should be noted that the modulational instability is always present for the modulation frequency within the range given by (9.20), although the integrated gain G is reduced by the fiber loss, as shown in (9.24). For this reason, the modulational instability is considered to deteriorate the capacity of coherent transmission systems [9.2, 9].

9.3 Induced Modulational Instability

When a small modulation is applied to an input signal, modulational instability can be induced if the side band frequency Ω falls within $\Omega < 2\Omega_\mathrm{m}$. This is called an induced modulational instability. If we make use of this induced modulational instability, it is possible to generate a train of soliton-like pulses with a repeti-

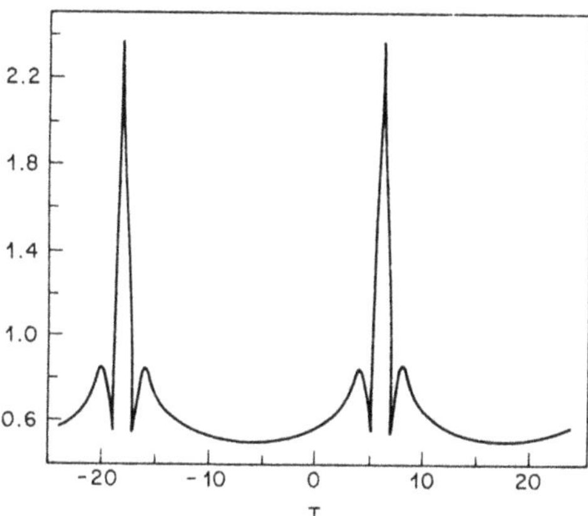

Fig. 9.3. Result of computer simulation of induced modulational instability [9.10]

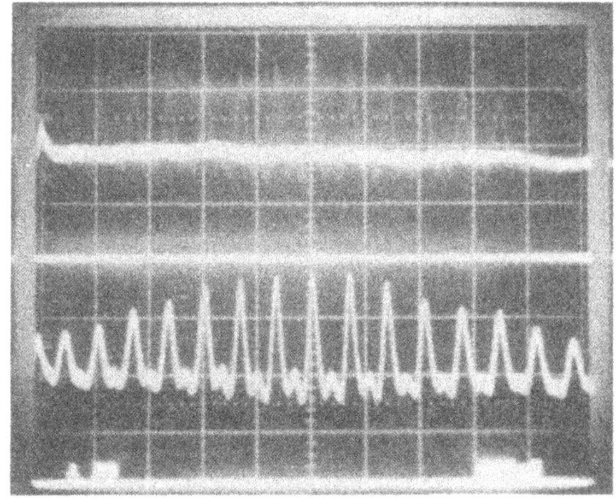

Fig. 9.4. Experimentally produced pulse train by induced modulational instability [9.11]

tion frequency determined by the inverse of the input modulation frequency Ω. Figure 9.3 shows the numerical evaluation of (9.1) for which the input signal is modulated by 20 % with a period of 24. One can see that a soliton-like pulse train can be generated from this initial condition [9.10].

A simple method of generating induced modulational instability experimentally is to inject two continuous light waves with different wavelengths into a fiber. Figure 9.4 is the autocorrelation output of the pulse train which was generated by this method [9.11]. For the input, the Nd : YAG laser with a peak power of 3 W, wavelength of 1.317 μm and a pulse width of 100 ps is used, together with an additional InGaAsP laser with only 0.2 mW and a frequency separation from the YAG laser of about 340 GHz. The fiber has a dispersion of -3.75 ps/(nm km) and a length of 1 km. It can be seen that almost 100 % modulation is generated at a distance of 1 km, even if the input modulation is only 0.017 %.

It should be noted that in the absence of fiber loss, the solution of (9.1) is repetitive and, therefore, this modulated pulse train regains the original shape of the continuous wave at a distance of about 2 km. Therefore, in order to produce the desired pulse train by this method, one should remove the pulse train at an appropriate distance. Induced modulational instability has an important application in the generation of a rapid pulse train with the desired duty cycle by simply providing two light waves into a fiber. However, it should be recognized that a wide band cw pump wave is required in order to suppress the induced Brillouin scattering.

Induced modulational instability can also be used to construct an ultrafast-all optical fiber switch with very large gain [9.12, 13] by utilizing the fact that an extremely small initial modulation can grow to the level of the pump amplitude along a rather short distance of fiber.

As another interesting application of modulational instability in a fiber, *Nakazawa* et al. [9.14] have recently demonstrated the feasibility of a new laser based on modulational instability.

10. Dark Solitons

In the wavelength range shorter than the zero group dispersion point where k'' is zero, the soliton solution does not exist. However, even in this range of wavelengths, the portion without light, which is produced by chopping a continuous light wave, is known to form a soliton. Such a soliton solution is often called a dark soliton. We present here the theoretical and experimental properties of a dark soliton.

10.1 Dark Soliton Solutions

When $k'' > 0$, (3.15) can be rewritten by using a new time variable T,

$$T = \frac{\tau}{(\lambda k'')^{1/2}} \quad , \tag{3.17'}$$

as

$$\mathrm{i}\frac{\partial q}{\partial Z} - \frac{1}{2}\frac{\partial^2 q}{\partial T^2} + |q|^2 q = 0 \quad . \tag{10.1}$$

In this expression, the definitions of q and Z are the same as those of (3.16) and (3.18), respectively. The dark soliton solution is obtained by introducing again the variables ϱ and σ,

$$q = \sqrt{\varrho}\,\mathrm{e}^{\mathrm{i}\sigma} \tag{10.2}$$

and by using the condition that ϱ becomes a function of T only (i.e., a stationary solution), in a manner similar to Sect. 3.3 [10.1]. Here, because of the opposite sign in the second term of the nonlinear Schrödinger equation (10.1), (3.26) reads

$$\left(\frac{d\varrho}{dT}\right)^2 = 4\varrho^3 + 8\Omega\varrho^2 + c_2\varrho - 4c_1^2 \quad . \tag{3.26'}$$

In this expression the double root appears for a larger value of ϱ than the single root. Consequently, unlike the case of the bright soliton of Sect. 3.3, c_1^2 can be positive and (3.26′) can be cast into the following form:

$$\left(\frac{d\varrho}{dT}\right)^2 = 4(\varrho - \varrho_0)^2(\varrho - \varrho_s) \quad , \tag{3.27'}$$

where ϱ_0, the double root, designates the asymptotic value of ϱ. Integrating (3.27′), we have,

$$\varrho = \varrho_0 \left[1 - a^2 \operatorname{sech}^2 (\sqrt{\varrho_0}\, aT)\right] \quad , \qquad a^2 = \frac{\varrho_0 - \varrho_s}{\varrho_0} \leq 1 \quad ,$$

$$\begin{aligned} \sigma &= \int \frac{c_1}{\varrho}\, dT + \Omega Z \\ &= \int \frac{\varrho_0^{3/2}(1-a^2)^{1/2}}{\varrho}\, dT - \frac{\varrho_0 (3-a^2)}{2} Z \\ &= \left[\varrho_0 (1-a^2)\right]^{1/2} T + \tan^{-1}\left[\frac{a}{\sqrt{1-a^2}} \tanh(\sqrt{\varrho_0}\, aT)\right] \\ &\quad - \frac{\varrho_0 (3-a^2)}{2} Z \quad , \end{aligned} \qquad (10.3)$$

$$c_1^2 = \varrho_0^3 (1 - a^2) \quad ,$$

$$\Omega = -\tfrac{1}{2}\, \varrho_0 (3 - a^2) \quad .$$

Unlike a bright soliton, a dark soliton has an additional new parameter, a, which designates the depth of modulation. We should also note the fact that at $T \to \pm\infty$, that phase of q changes. Such a soliton is called a topological soliton while the bright soliton, which has no phase change at $T \to \pm\infty$, is called a nontopological soliton. When $a = 1$, the depth approaches 0 and the solution becomes

$$q = \sqrt{\varrho_0} \tanh(\sqrt{\varrho_0}\, T) \quad . \qquad (10.4)$$

As in the case of a bright soliton solution, a general dark soliton solution can be obtained by a Galilei transformation of (10.4) and is given by

$$\varrho' = \varrho_0 \left\{1 - a^2 \operatorname{sech}^2 [\sqrt{\varrho_0}\, a(T - UZ)]\right\} \qquad (10.5)$$

$$\sigma' = \sigma + uT - \tfrac{1}{2} u^2 Z \quad .$$

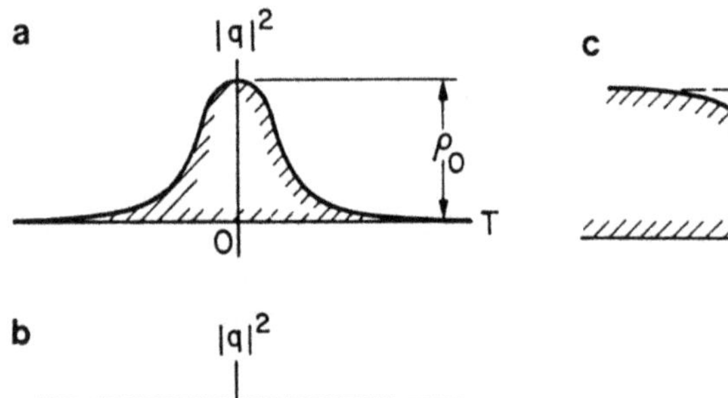

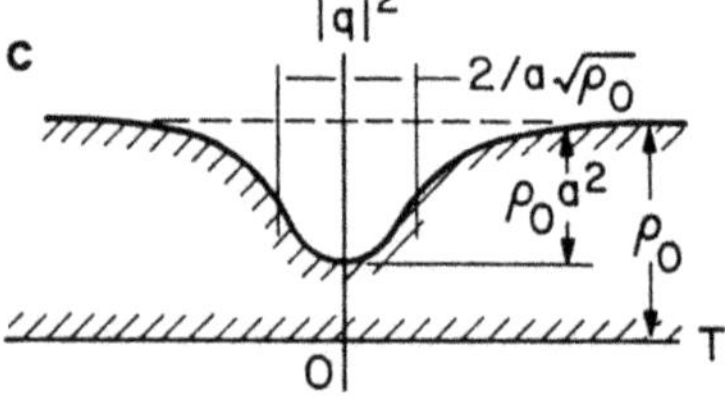

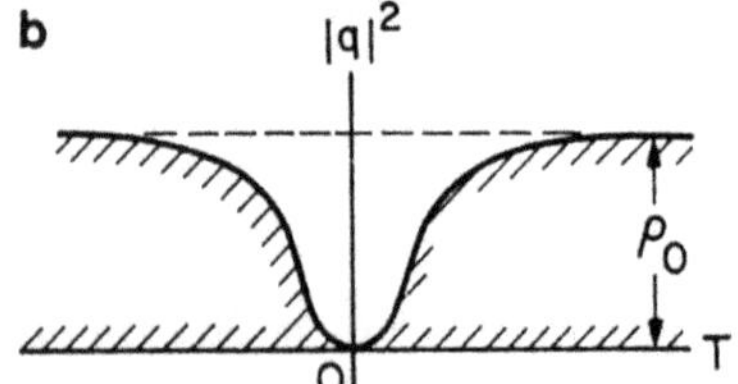

Fig. 10.1. Bright (**a**) and dark solitons (**b**) and (**c**)

The two types of dark solitons are shown in Fig. 10.1 and compared with a bright soliton. The inverse scattering method for (10.1) was also discovered by *Zakharov* and *Shabat* [10.2], and dark solitons are shown to correspond to the eigenvalues of Dirac-type equations similar to (4.6) and (4.7).

10.2 Experimental Results of Dark Solitons

Dark solitons in fibers were first observed by *Emplit* et al. [10.3] and *Krökel* et al. [10.4] independently by transmitting a light wave in the normal dispersion region of a fiber. Figure 10.2 shows the experimental result obtained by Krökel et al. They used a YAG laser with an output of 100 ps in which 0.3 ps holes were produced by a modulator into a 10 m single-mode optical fiber. The output signals were measured using the autocorrelation technique for various input power levels and the results were compared with the numerical solution of the nonlinear Schrödinger equation (10.1), as shown in Fig. 10.3. As seen from Fig. 10.2, when the input power levels were increased from 0.2 W (a) to 2 W (b), to 9 W (c) and

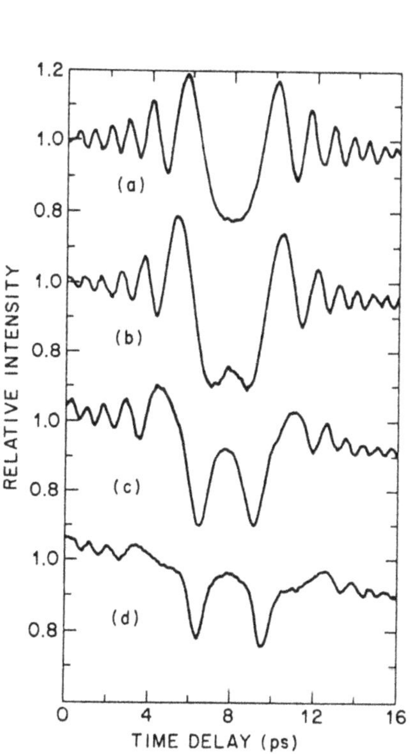

Fig. 10.2. Experimental results of formation of dark solitons [10.4]

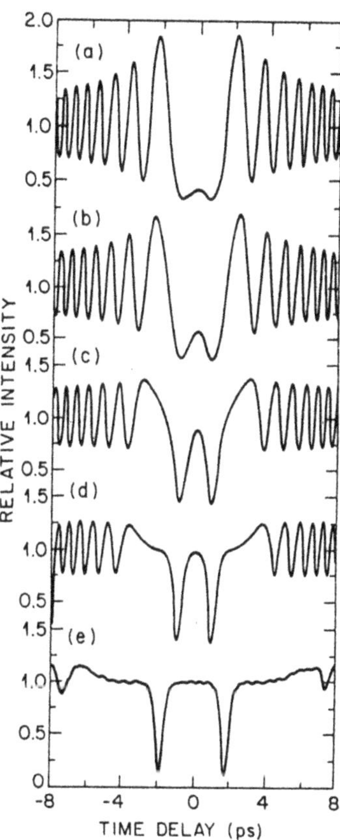

Fig. 10.3. Numerical results of formation of dark solitons [10.4]

to 20 W (d), the linear dispersive response at (a) was gradually modified to form a pair of dark solitons at 20 W power level, as shown in (d).

The numerical results of Fig. 10.3 are the outputs for 0.2 W (a), 4 W (b), 18 W (c) and 40 W (d), while (e) is for 40 W with a distance of 20 m.

Good agreement between the experimental and numerical results can be seen. The production of a pair of dark solitons is a consequence of the fact that the input pulse does not have the phase jump needed for a dark soliton solution, as shown in (10.3). If the input pulse has the same phase at $T \to \pm\infty$, a pair of dark solitons appears, so that the one dark soliton is reversed by the other dark soliton.

Since a dark soliton is a topological soliton, in order to form a single dark soliton one should construct a dark pulse with an appropriate phase change. Such an experiment was recently performed by *Weiner* et al. [10.5] by reversing the phase at the middle of a few-picosecond pulse.

Figure 10.4 shows their experimental results (dotted lines) where (a) is the input dark pulse formed in the middle of a bright pulse while (b)–(e) are output pulses emerging from the normal dispersion fiber for peak powers of (b) 1.5, (e) 52.5, (d) 150, and (e) 300 mW. The solid lines show theoretical results obtained by numerically integrating the nonlinear Schrödinger equation (10.1). One can see the excellent agreement between observations and theoretical calculations in the narrowing of the dark soliton with increasing power level. Unlike a bright soliton, no higher order solitons exist. The dark soliton simply becomes narrower as the power level is increased.

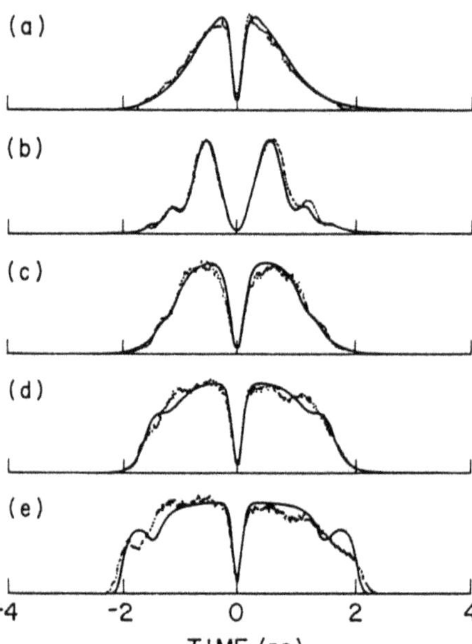

Fig. 10.4. Experimental results (*dotted line*) of dark soliton propagation by *Weiner* et al. [10.5] for increasing power levels (b)–(e). (a) Input pulse shape; solid line is the theoretical result

11. Summary

It is approximately 15 years since the first theory of optical solitons in fibers was published. During this period, not only was the experimental verification of the existence of solitons (both bright and dark) demonstrated, but also various properties of the soliton derivable from the result of the inverse scattering theory were identified. The model of the nonlinear Schrödinger equation for describing optical solitons in a glass fiber has been established.

Recently, the effects of higher order terms on soliton transmission, in particular, those which originate from stimulated Raman scattering in fibers, have also been experimentally identified. In addition, soliton processes have been applied to a soliton laser, or soliton compression. However, what is perhaps the most important application, i.e., the application to high speed optical transmission systems, has not yet materialized. One reason for this is the unavailability of a high power infra-red solid state laser. However, the recent demonstration of the transmission of an optical soliton amplified using the Raman process in a fiber over a distance of 6000 kms promises a potential application in an all-optical high speed transmission system.

At present, an optical transmission system of approximately 2 Gbit/s is being developed and one beyond 10 Gbit/s is being investigated. However, repeaters for a high speed transmission system become increasingly expensive due to the necessity for high speed electronic to photonic coupling. Because of this, a system which has 10 times the capacity of the present system is inconceivable by a simple extension of the present linear system. Consequently, the optical soliton is regarded as an important alternative for the next generation of high speed transmission sytems.

The concept of the soliton involves a large number of interesting problems in applied mathematics since it is an exact analytical solution of a nonlinear partial differential equation. In natural science, the soliton concept is applied to surface and internal water waves, waves in the air and the formation of a star. In physics, the concept of solitons is applied to plasmas and condensed matter, as well as to particle physics. However, to the knowledge of the author, only the theory of optical solitons described by the nonlinear Schrödinger equation has produced perfect agreement between theory and experiment and shown solitons to have potential applications in many other domains of optic areas. It is expected that research on optical solitons will develop further theoretically, experimentally and in the quest for applications.

There are many interesting problems of optical solitons in fibers which are not treated in this book but deserve careful attention. One is the effect of polarization.

The rigorous derivation of the nonlinear Schrödinger equation in Sect.3.4 for a guided mode assumes existence of a single eigenfunction. This assumption is justified for a polarization preserving fiber but not for a standard fiber where modes with two different polarizations are mixed. Due to small inhomogeneities in the fiber the modes with two different polarizations may acquire different group velocities as they propagate over an extended distance. As a result a soliton which contains modes with two polarizations faces additional dispersion called polarization dispersion [11.1]. This phenomenon may be analyzed by a model of nonlinearly coupled nonlinear Schrödinger equations [11.2,3] whose general properties are not yet known. The polarization dispersion may give rise to most serious effects in the long distance transmission of optical solitons.

Another interesting phenomenon is the effect of repeated perturbations in terms of amplification, loss, and/or fiber inhomogeneity have on the soliton propagation. For linear and periodic gain, stable transmission has been demonstrated both in experiments [11.4] and in simulations [11.5]. However if the perturbation is stochastic, the soliton parameters (such as its amplitude and velocity) may face diffusion [11.6,7] and can become chaotic if the perturbation is of parametric type [11.8] — even if the shape of solitons themselves remains intact.

One interesting phenomenon in this regard is that there exists a class of nonlinear perturbations which can lead to a fix point in the chaotic motion of the soliton parameters [11.9]. This property may be used to lock soliton velocities (and amplitudes) which otherwise may jitter due to perturbations.

A further point of note is the discover of auto-solitons, i.e., self-organized solitons [11.10,11], in the presence of a nonlinear perturbation, in a fiber with nonlinearly saturated gain (such as a semiconductor laser amplifier) where a new soliton emerges with parameters determined by the perturbation quantities. These phenomena arise from nonlinear perturbations and are unique intrinsic properties of solitons being nonlinear objects. They may be used to construct advanced repeaters of a long distance soliton transmission system.

General References

Abdullaev, F.Kh., Damanian, S.A., Khabibullaev, P.K.: *Optical Solitons*, ed. by V.G. Baryachtar (Fan,Tashkent 1987)

Anderson, D., Lisak, M.: "Asymptotic linear dispersion of optical pulses in the presence of fiber nonlinearity and loss", Opt. Lett. **10**, 390 (1985)

Andrushko, L.M., Karplyuk, K.S., Ostrovskii, S.B.: "On the propagation of solitons in coupled optical fibers", Sov. J. Commun. Techol. Electron **32**, 161 (1987)

Crosignani, B., Cutolo, A., DiPorto, P.: "Coupled-mode theory of nonlinear propagation in multimode and single mode fibers: envelope solitons and self-confinement", J. Opt. Soc. Am. **72**, 1136 (1982)

Mitschke, F.M.: "Solitons in optical fibers: light pulse with surprising properties", Laser Optoelektron **19**, 393 (1987)

Pentrun'kin, V.Yu., Selishchev, A.V., Sysuev, A.V., Shcherbakov, A.S., Kalinin, M.I.: "Soliton mode of light pulse propagation in single-mode optical fibers and the question of its experimental realization", Pis'ma v Zh. Tekh. Fiz. **12**, 988 (1986) [Trans. Sov. Tech. Phys. Lett. **12**, 408 (1986)]

Sisakyan, I.N., Shvartsburg, A.B.: "Nonlinear dynamics of picosecond pulses in fiber-optic waveguides", Kvantovaya Elektron. **11**, 1703 (1984) [Trans: Sov. J. Quant. Electron. **14**, 1146 (1984)]

Weisi, T.: "Optical soliton propagation in ideal monomode fiber", Chin. J. Lasers **14**, 625 (1987)

General References

Abdullaev, F.Kh., Darmanyan, S.A., Khabibullaev, P.K.: Optical Solitons [illegible]

[illegible]

References

Chapter 1

1.1 N.J. Zabusky, M.D. Kruskal: Phys. Rev. Lett. **15**, 240 (1965)
1.2 D.J. Korteweg, G. deVries: Phil. Mag. Ser. 5, **39**, 422 (1895)
1.3 C.S. Gardner, J.M. Greene, M.D. Kruskal, R.M. Miura: Phys. Rev. Lett. **19**, 1095 (1967)
1.4 T. Taniuti, H. Washimi: Phys. Rev. Lett. **21**, 209 (1968)
1.5 A. Hasegawa: Phys. Fluids **24**, 1165 (1970)
1.6 R.Y. Chiao, E. Garmire, C.H. Townes: Phys. Rev. Lett. **13**, 479 (1964)
1.7 A. Hasegawa, F.D. Tappert: Appl. Phys. Lett. **23**, 142 (1973)
1.8 A. Hasegawa, F.D. Tappert: Appl. Phys. Lett. **23**, 171 (1973)
1.9 V.E. Zakharov, A.B. Shabat: Zh. Eksp. i Teor. Fiz. **61**, 118 (1971) [Trans. Soviet. Phys. - JETP **34**, 62 (1972)]
1.10 S.A. Akhmanov, A.P. Sukhorukov, R.V. Khokhlov: Sov. Phys. USPEKHI, **93**, 609 (1968)
1.11 A.M. Prokhorov: Bull. Acad. Sci. USSR Phys. Ser. **47**, 1 (1983)
1.12 V.Sohor, T.T. Tam, F. Varga: Bull. Acad. Sci. USSR Phys. Ser. **47**, 71 (1983)
1.13 L.F. Mollenauer, R.H. Stolen, J.P. Gorden: Phys. Rev. Lett. **45**, 1095 (1980)
1.14 A. Hasegawa, W.F. Brinkmann: IEEE J. Quant. Elect. QE-**16**, 694 (1980)
1.15 A. Anderson, M. Lisak: Opt. Lett. **9**, 463 (1984)
1.16 B. Hermansson, D. Yerick: Opt. Comm. **52**, 99 (1984)
1.17 K. Tai, A. Hasegawa, A. Tomita: Phys. Rev. Lett. **59**, 135 (1986)
P. Emplit, J.P. Hamaide, F. Reynaud, C. Froely, A. Barthelemy: Opt. Comm. **62**, 374 (1987)
1.18 D. Krökel, N.J. Halas, G. Giuliani, D. Grischkowsky: Phys. Rev. Lett. **60**, 29 (1988)
1.19 A. Hasegawa: Appl. Opt. **23**, 3302 (1984)
1.20 L.F. Mollenauer, K. Smith: Opt. Lett. **13**, 675 (1988) and to be published
1.21 L.F. Mollenauer, R.H. Stolen, J.P. Gordon, W.J. Tomlinson: Opt. Lett. **8**, 289 (1983)
1.22 L.F. Mollenauer: Phil. Trans. R. Soc. Lond. **A315**, 437 (1985)
1.23 F.M. Mitschke, L.F. Mollenauer: Opt. Lett. **11**, 657 (1986)
1.24 J.P. Gordon: Opt. Lett. **11**, 662 (1986)
1.25 Y. Kodama, A. Hasegawa: IEEE J. Quant. Elect. QE-**23**, 510 (1987)
1.26 P. Beuad, W. Hodel, B. Zysset, H.P. Weber: IEEE J. Quantum. Electron. **QE-23**, 1938 (1987)

1.27 K. Tai, A. Hasegawa, N. Bekki: Opt. Lett. **13**, 392 and 937 (1988)
1.28 A. Hasegawa, Y. Kodama: Proc. IEEE **69**, 1145 (1981)
1.29 H.G. Unger: Proc. Int. Conf. on Communications **1**, 153 (1984)
1.30 J.E. Midwinter: Telephony **207**, 56 (1984)
1.31 N.J. Doran, K.J.Blow: IEEE J. Quant. Elect.QE-**19**, 1883 (1983)

Chapter 2

2.1 C.S. Gardner, J.M. Greene, M.D. Kruskal, R.M. Miura: Phys. Rev. Lett. **19**, 1095 (1967)
2.2 P.D. Lax: Commun. Pure and Appl. Math. **10**, 537 (1957) and **21**, 467 (1968)
2.3 V.E. Zakharov, A.B. Shabat: Zh. Eksp. i Teor. Fiz **61**, 118 (1971) [Trans. Sov. Phys. JETP **34**, 62 (1972)]

Chapter 3

3.1 A. Hasegawa, Y. Kodama: Proc. IEEE **69**, 1145 (1981)
3.2 S.A. Akhmanov, A.P. Sukhorukov, R.V. Khoklov: Usp. Fiz. Nauk **93**, 19 (1967) [Trans. Sov. Phys. USPEKHI **93**, 609 (1968)]
3.3 V.E. Zakharov, A.B. Shabat: Zh. Eksp. i Teor. Fiz. **61**, 118 (1971) [Trans. Soviet Phys. - JEPT **34**, 62 (1972)]
3.4 A. Hasegawa, F.D. Tappert: Appl. Phys. Lett. **23**, 171 (1973)
3.5 Y. Kodama: J. Stat. Phys. **39**, 597 (1985)
3.6 Y. Kodama, A. Hasegawa: IEEE J. Quant. Elect. QE-**23**, 510 (1987)
3.7 T. Taniuti: Prog. Theor. Phys. (Japan), Suppl. **55**, 1 (1974)
3.8 A. Hasegawa, F. Tappert: Appl. Phys. Lett. **23**, 142 (1973)
3.9 J.P. Gordon: Opt. Lett. **11**, 662 (1986)

Chapter 4

4.1 V.E. Zakharov, A.B. Shabat: Zh. Eksp. i Teor. Fiz. **61**, 118 (1971) [Trans. Soviet Phys. - JETP **34**, 62 (1972)]
4.2 C.S. Gardner, J.M. Greene, M.D. Kruskal, R.M. Miura: Phys. Rev. Lett. **19**, 1095 (1967)
4.3 P.D. Lax: Commun. Pure and Appl. Math. **10**, 537 (1957) and **21**, 467 (1968)
4.4 J. Satsuma, N. Yajima: Suppl. Prog. Theor. Phys. (Japan) **55**, 284 (1974)
4.5 Y. Kodama, A. Hasegawa: IEEE J. Quant. Elect. QE-**23**, 510 (1987)

Chapter 5

5.1 A. Hasegawa, Y. Kodama: Proc. IEEE **69**, 1145 (1981)
5.2 E.M. Dianov, N.S. Nikonova, V.N. Serkin: Kvantovaya Elektron **13**, 331 (1986) [Trans. Sov. J. Quant. Electron **16**, 219 (1986)]
5.3 K. Tajima, K. Washio: Oyo Buturi **55**, 654 (1986)
5.4 G.N. Burlak, N. Ya Kotsarenko: Pis'ma v Zh. Tekh. Fiz. **10**, 674 (1984) [Trans. Sov. Tech. Phys. Lett. **10**, 284 (1984)]

5.5 F.Z. El-Halafawy, E.-S.A. El-Badawy, M.W. El-Gammal, M.H.A. Hasan: IEEE Trans. Instrum. Meas. IM **36**, 543 (1987)
5.6 G.N. Burlak, V.V. Grimal'skii, N. Ya Kotsarenko: Izv. vuz Radiofig. **29**, 1259 (1986) [Trans. Radiophys. and Quant. Elect. **29**, 959 (1986)]
5.7 A.I. Maimistov, E.A. Manykin, Yu.M. Sklyarov: Kvantovaya Elektron. **13**, 2243 (1986) [Trans. Sov. J. Quant. Electron **16**, 1480 (1986)]
5.8 E.M. Dianov, A.M. Prokhorov, V.N. Serkin: Dokl. Akad. Nank SSSR **273**, 1112 (1983) [Trans. Sov. Phys. Dokl. **28**, 1036 (1983)]
5.9 K.J. Blow, N.J. Doran, D. Wood: Opt. Lett. **12**, 202 (1987)
5.10 C.R. Menyuk: Opt. Lett. **12**, 614 (1987)
5.11 D.N. Christodoulides: Phys. Lett. A **132**, 451 (1988)
5.12 D. Anderson, M. Lišak, T. Reichel: J. Opt. Soc. Am. B. Opt. Phys. **5**, 207 (1988)
5.13 P.K.A. Way, C.R. Menyuk, H.H. Chen, Y.C.Lee: Opt. Lett. **12**, 628 (1987)
5.14 G.P. Agrawal, M.J. Potasek: Phys. Rev. A **33**, 1765 (1986)
5.15 L.F. Mollenauer, R.H. Stolen, J.P. Gordon: Phys. Rev. Lett. **45**, 1095 (1980)
5.16 R.H. Stolen, L.F. Mollenauer, W.J. Tomlinson: Opt. Lett. **8**, 186 (1983)
5.17 B. Zysset, P. Beaud, W. Hodel: Appl. Phys. Lett. **50**, 1027 (1987)
5.18 F.W. Wise, I.A. Walmsley, C.L. Tang: Opt. Lett. **13**, 129 (1988)
5.19 K. Watsuki, A. Takada, M. Saruwatari: Proc. IQEC, PD-14, Tokyo (1988)
5.20 C. Desem and P.L. Chu: Opt. Lett. **11**, 248 (1986)

Chapter 6

6.1 A. Hasegawa, Y. Kodama: Opt. Lett. **7**, 285 (1982)
6.2 Y. Kodama, A. Hasegawa: Opt. Lett. **7**, 339 (1982)
6.3 Y. Kodama, A. Hasegawa: Opt. Lett. **8**, 342 (1983)
6.4 C.R. Menyuk, H.H. Chen, Y.C. Lee: Opt. Lett. 451 (1985)
6.5 K. Tajima: Opt. Lett. 54 (1987)
6.6 K.J. Blow, N.J. Doran, D. Wood: J. Opt. Soc. Soc. Am. B. Opt. Phys. **5**, 381 (1988)
6.7 A. Hasegawa: Opt. Lett. **8**, 650 (1983)
6.8 V.A. Vysloukh, V.N. Serkin: Pis'ma v Zh. Eksq. and Teor. **38**, 170 (1983) [Trans. JETP Lett. **38**, 199 (1983)]
6.9 L.F. Mollenauer, R.H. Stolen: Opt. Lett. **10**, 229 (1985)
6.10 E.M. Dianov, Z.S. Nikonova, A.M. Prokhorov, V.N. Serkin: Dokl. Akad. Nauk SSSR, **283**, 1342 (1985) [Trans. Sov. Phys. Dokl. **30**, 689 (1985)]
6.11 V.M. Mitev, L.M. Kovachev: Opt. Comm. **63**, 421 (1987)
6.12 R.J. Mears, L. Reekie, I.M. Jauncey, D.N. Payne: Elect. Lett. **23**, 1026 (1987)
6.13 E. Desurvire, J.R. Simpson, P.C. Becker: Opt. Lett. **12**, 888 (1987)
6.14 M. Nakazawa, Y. Kimura, K. Suzuki: Elect. Lett. **25**, 199 (1989)
6.15 M. Nakazawa, Y. Kimura, K. Suzuki: Appl. Phys. Lett. **54**, 295 (1989)
6.16 M. Nakazawa, Y. Kimura, K. Suzuki: Opt. Lett., to be published
6.17 L.F. Mollenauer, J.P. Gordon, M.N. Islam: IEEE J. Quant. Elect. **QE-22**, 157 (1986)

6.18 R.H. Stolen, E.P. Ippen: Appl. Phys. Lett. **22**, 276 (1973)
6.19 A. Hasegawa: Appl. Opt. **23**, 3302 (1984)
6.20 V.I. Karpman, V.V. Solov'ev: Physica, **3D**, 487 (1981)
6.21 J.P. Gordon: Opt. Lett. **8**, 596 (1983)
6.22 D. Anderson, M. Lisak: Phys. Rev. A **32**, 2270 (1985)
6.23 E. Shiojiri, Y. Fujii: Appl. Opt. **24**, 358 (1985)
6.24 F.M. Mitschke, L.F. Mollenauer: Opt. Lett. **12**, 355 (1987)
6.25 C. Desem, P.L. Chu: Opt. Lett. **12**, 349 (1987)
6.26 J.P. Gordon, H.A. Haus: Opt. Lett. **11**, 665 (1986)
6.27 L.F. Mollenauer, K. Smith: Opt. Lett. **13**, 675 (1988)
6.28 L.F. Mollenauer, K. Smith: to be published
6.29 S. Chi, S. Wen: Proc. SPIE Int. Soc. Opt. Eng. **613**, 119 (1986)
6.30 A.S. Gouveia-Neto, A.S.L. Gomes, J.R. Taylor, K.J. Blow: J. Opt. Soc. Am. B. Opt. Phys. **5**, 799 (1988)
6.31 A.S. Gouveia-Neto, A.S.L. Gomes, J.R. Taylor: Opt. Lett. **12**, 1035 (1987)
6.32 G.M. Dianov, A.B. Grudinin, D.V. Khaidarov, D.V. Korobkin, A.M. Prokhorov, V.N. Serkin: 13th European Conf. on Opt. Comm. Tech. Digest **1**, 211 (1987)
6.33 K. Iwatsuki, S. Nishi, M. Saruwatari, M. Shimizu: IOOC 1989 Kobe, Japan Tech. Digest **5**, paper 20PDA-1 (1989)

Chapter 7

7.1 L.F. Mollenauer, R.H. Stolen, J.P. Gordon, W.J. Tomlinson: Opt. Lett. **8**, 289 (1983)
7.2 H.A. Haus, M.N. Islam: IEEE J. Quant. Elect. **QE-21**, 1172 (1985)
7.3 P. Berg, F. If, P.L. Christiansen, D. Skovgaard: Phys. Rev. A **35**, 4167 (1987)
7.4 P.A. Belanger: J. Opt. Soc. Am. B. Opt. Phys. **5**, 793 (1988)
7.5 F.M. Mitschke, L.F. Mollenauer: IEEE J. Quant. Elect. **QE-22**, 2242 (1986)
7.6 K.J. Blow, D. Wood: IEEE J. Quant. Elect. **QE-22**, 1109 (1986)
7.7 E.M. Dianov, A.M. Prokhorov, V.N. Serkin: Opt. Lett. **11**, 168 (1986)
7.8 J.D. Kafka, T. Baer: Hyperfine Interact. **37**, 291 (1987)
7.9 A.S. Gouveia, A.S.L. Gomes, J.R. Taylor, B.J. Ainslie, S.P. Craig: Opt. Lett. **12**, 927 (1987)
7.10 J.D. Kafka, T. Baer: Opt. Lett. **12**, 181 (1987)
7.11 M.N. Islam, L.F. Mollenauer, R.H. Stolen: *Ultrafast Phenomena*, ed. by G.R. Fleming, A. Sigman (Springer, Berlin, NewYork 1986) p.46
7.12 K. Tai, A. Hasegawa, N. Bekki: Opt. Lett. **13**, 392 and 937 (1988)
7.13 L.F. Mollenauer: Phil. Trans. R. Soc. Lond. **A315**, 437 (1985)
7.14 F.M. Mitschke, L.F. Mollenauer: Opt. Lett. **12**, 407 (1987)
7.15 L.F. Mollenauer, R.H. Stolen, J.P. Gordon, W.J. Tomlinson: Opt. Lett. **8**, 289 (1983)
7.16 J.A. Valdmanis, P.L.Fork: IEEE J. Quant. Elect. **QE-22**, 112 (1986)
7.17 D. Mestaagh: Appl. Opt. **26**, 5234 (1987)

7.18 E.M. Dianov, Z.S. Nikonova, A.M. Prokhorov, V.N. Serkin: Pis'ma v Zh. Tekh. Fiz. **12**, 756 (1986) [Trans. Sov. Tech. Phys. Lett. **12**, 311 (1986)]
7.19 C.V. Shank, R.L. Fork, R. Yen, R.H. Stolen, W.J. Tomlinson: Appl. Phys. Lett. **40**, 76 (1982)
7.20 W.J. Tomlinson, R.H. Stolen, C.V. Shank: J. Opt. Soc. Am. B. **1**, 139 (1984)
7.21 K. Tai, A. Tomita: Appl. Phys. Lett. **48**, 1033 (1986)
7.22 A.S. Gouveia-Neto, A.S.L. Gomes, J.R. Taylor: Opt. Lett. **12**, 395 (1987)
7.23 A.S. Gouveia, A.S.L. Gomes, J.R. Taylor: IEEE Quant. Elect. **QE-23**, 1193 (1987)
7.24 W. Hodel, H.P. Weber: Opt. Lett. **12**, 924 (1987)
7.25 A.S. Gouveia, A.S.L. Gomes, J.R. Taylor: IEEE Quant. Elect. **QE-24**, 332 (1988)
7.26 R.H. Enns, S.S. Rangnekar: Opt. Lett. **12**, 108 (1987)
7.27 R.H. Enns, S.S. Rangnekar: IEEE J. Quant. Elect. **QE-23**, 1199 (1987)
7.28 N.J. Doran, D. Wood: J. Opt. Soc. Am. B. Opt. Phys. **4**, 1843 (1987)

Chapter 8

8.1 Y. Kodama, A. Hasegawa: IEEE J. QE. **QE-23**, 510 (1987)
8.2 N. Tzoar, M. Jain: Phys. Rev. A **23**, 1266 (1981)
8.3 D. Yevich, B. Hermansson: Opt. Commun. **47**, 101 (1983)
8.4 D.N. Christodoulides, R.I.Joseph: Appl. Phys. Lett. **47**, 76 (1985)
8.5 E.A. Golovchenko, E.M. Dianov, A.M. Prokhorov, V.N. Serkin: Pis'ma Zh. Eksp. Fiz. **42**, 74 (1985) [Trans. JETP Lett. **42**, 87 (1985)]
8.6 K. Ohkuma, Y.H. Ichikawa, Y. Abe: Opt. Lett. **12**, 516 (1987)
8.7 V.N. Serkin: Pis'ma Zh. Tekh. Fiz. **13**, 772 (1987) [Trans. Sov. Tech. Phys. Lett. **13**, 320 (1987)]
8.8 F.M. Mitschke, L.F. Mollenauer: Opt. Lett. **11**, 659 (1986)
8.9 J.P. Gordon: Opt. Lett. **11**, 662 (1986)
8.10 W. Hodel, H.P. Weber: Opt. Lett. **12**, 924 (1987)
8.11 Y. Kodama, K. Nozaki: Opt. Lett. **12**, 1038 (1987)
8.12 K. Tai, A. Hasegawa, N. Bekki: Opt. Lett. **13**, 392 (1988); **13**, 937 (1988)
8.13 A.B. Grudinin, E.M. Dianov, D.V. Korobkin, A.M. Prokhorov, V.N. Serkin, D.V. Khaidarov: Pis'ma Zh. Eksp. Teor. Fiz. **46**, 175 (1987) [Trans. JETP Lett. **46**, 221 (1987)]
8.14 P. Beaud, W. Hodel, B. Zysset, H.P. Weber: IEEE J. Quant. Elect. **QE-23**, 1938 (1987)
8.15 Y. Kodama: J. Stat. Phys. **39**, 597 (1985)
8.16 Y. Kodama: Phys. Lett. **107A**, 245 (1985)

Chapter 9

9.1 A. Hasegawa, W.F. Brinkman: IEEE J. Quant. Elect. **QE-16**, 694 (1980)
9.2 D. Anderson, M. Lisak: Opt. Lett. **9**, 463 (1984)
9.3 B. Hermansson, D. Yevich: Opt. Comm. **52**, 99 (1984)

9.4 K. Tajima: J. Lightwave Tech. **LT-4**, 900 (1986)
9.5 P.K. Shukla, J.J. Rasmussen: Opt. Lett. **11**, 171 (1986)
9.6 M.J. Potasek, G.P. Agrawal: Phys. Rev. **A36**, 3862 (1987)
9.7 G.P. Agrawal: Phys. Rev. Lett. **59**, 880 (1987)
9.8 K. Tai, A. Hasegawa, A. Tomita: Phys. Rev. Lett. **59**, 135 (1986)
9.9 A. Hasegawa, K. Tai: Opt. Lett. **14**, to be published
9.10 A. Hasegawa: Opt. Lett. **9**, 288 (1984)
9.11 K. Tai, A. Tomita, J.L. Jewell, A. Hasegawa: Appl. Phys. Lett. **49**, 236 (1986)
9.12 M.N. Islam, S.P. Dijaili, J.P. Gordon: Opt. Lett. **13**, 518 (1988)
9.13 C.E. Soccolich, M.N. Islam: Opt. Lett. **14**, to be published
9.14 M. Nakazawa, K. Suzuki, H.A. Haus: IEEE J. Quant. Elect., to be published

Chapter 10

10.1 A. Hasegawa, F.D. Tappert: Appl. Phys. Lett. **23**, 171 (1973)
10.2 V.E. Zakharov, A.B. Shabat: Zh. Eksp. Teor. Fiz. **64**, 1627 (1973) [Trans. Sov. Phys. JETP **37**, 823 (1974)]
10.3 P. Emplit, J.P. Hamaide, F. Reynaud, C. Froehly, A. Barthelemy: Opt. Comm. **62**, 374 (1987)
10.4 D. Krökel, N.J. Halas, G. Giuliani, D. Grischkowsky: Phys. Rev. Lett. **60**, 29 (1988)
10.5 A.M. Weiner, J.P. Heritage, R.J. Hawkins, R.N. Thurston, E.M. Kirschner, D.E. Leaird, W.J. Tomlinson: Phys. Rev. Lett. **21**, 2445 (1988)

Chapter 11

11.1 I.P. Kaminow: IEEE J. Quantum Electron. **QE-17**, 15 (1981)
11.2 C.R. Menyuk: Opt. Lett. **12**, 614 (1987)
11.3 C.R. Menyuk: J. Opt. Soc. Am. B **5**, 392 (1988)
11.4 L.F. Mollenauer, K. Smith: Opt. Lett. **13**, 675 (1988)
11.5 E. Desurvire, J.R. Simpson, P.C. Becker: Opt. Lett. **12**, 888 (1987)
11.6 Y. Kodama, A. Hasegawa: Opt. Lett. **8**, 342 (1983)
11.7 F.G. Bass, Yu.S. Kivshar, V.V. Konotop, G.M. Pritula: Opt. Commun. **70**, 309 (1989)
11.8 K. Nozaki, N. Bekki: Phys. Rev. Lett. **50**, 1226 (1983); J. Phys. Soc. Jpn. **54**, 2363 (1985)
11.9 Y. Kodama: J. Stat. Phys. **39**, 597 (1985)
11.10 V.S. Grigoryán: JETP Lett. **44**, 575 (1986)
11.11 V.S. Grigoryán, A.I. Maimistov, Yu.M. Sklyarov: Soc. Phys. JETP **67**, 530 (1988)

Subject Index

www.ingramcontent.com/pod-product-compliance
Ingram Content Group UK Ltd.
Pitfield, Milton Keynes, MK11 3LW, UK
UKHW021821190726
13853UKWH00003B/1106

* 9 7 8 3 6 6 2 0 9 1 1 4 2 *